BRITISH No.3

MUL ...TS

SIXTEENTH EDITION
2003

The complete guide to all Diesel Multiple
Units which operate on Network Rail

Peter Fox & Robert Pritchard

UPDATED TO 2004 EDITION 1-1.04

ISBN 1 902336 28 3

© 2002. Platform 5 Publishing Ltd., 3 Wyvern House, Sark Road, Sheffield,
S2 4HG, England.

Pocket

CONTENTS

PROVISION OF INFORMATION

This book has been compiled with care to be as accurate as possible, but in some cases official information is not available and the publisher cannot be held responsible for any errors or omissions. We would like to thank the companies and individuals which have been co-operative in supplying information to us. The authors of this series of books will be pleased to receive notification from readers of any inaccuracies readers may find in the series, and also notification of any additional information to supplement our records and thus enhance future editions. Please send comments to:

Robert Pritchard, Platform 5 Publishing Ltd., Wyvern House, Sark Road, Sheffield, S2 4HG, England.
Tel: 0114 255 2625 Fax: 0114 255 2471
e-mail: platfive@platfive.freeserve.co.uk

Both the author and the staff of Platform 5 regret they are unable to answer specific queries regarding locomotives and rolling stock.

This book is updated to 25 November 2002.

UPDATES

An update to all the books in the *British Railways Pocket Book* series is published every month in the Platform 5 magazine, **entrain**, which contains news and rolling stock information on the railways of Britain and Ireland. For further details of **entrain**, please see the advertisement on the back cover of this book.

BRITAIN'S RAILWAY SYSTEM

INFRASTRUCTURE & OPERATION

Britain's national railway infrastructure is now owned by a "not for dividend" company, Network Rail, following the demise of Railtrack. Many stations and maintenance depots are leased to and operated by Train Operating Companies (TOCs), but some larger stations remain under Network Rail control. The only exception is the infrastructure on the Isle of Wight, which is nationally owned and is leased to the Island Line franchisee.

Trains are operated by TOCs over Network Rail, regulated by access agreements between the parties involved. In general, TOCs are responsible for the provision and maintenance of the locomotives, rolling stock and staff necessary for the direct operation of services, whilst Network Rail is responsible for the provision and maintenance of the infrastructure and also for staff needed to regulate the operation of services.

DOMESTIC PASSENGER TRAIN OPERATORS

The large majority of passenger trains are operated by the TOCs on fixed term franchises. Franchise expiry dates are shown in parentheses in the list of franchisees below:

Franchise	Franchisee	Trading Name
Anglia Railways	GB Railways plc. (until 4 April 2004)	Anglia Railways
Central Trains	National Express Group plc (until 1 April 2004)	Central Trains
Chiltern Railways	M40 Trains Ltd. (until December 2021)	Chiltern Railways
Cross Country	Virgin Rail Group Ltd.* (until March 2012)	Virgin Trains
Gatwick Express	National Express Group plc (until 27 April 2011)	Gatwick Express
Great Eastern Railway	First Group plc (until 4 April 2004)	First Great Eastern
Great Western Trains	First Group plc (until 3 February 2006)	First Great Western
InterCity East Coast	GNER Holdings Ltd. (until 4 April 2005)	Great North Eastern Railway
InterCity West Coast	Virgin Rail Group Ltd.* (until 8 March 2012)	Virgin Trains
Island Line	Stagecoach Holdings PLC (until February 2007)	Island Line
LTS Rail	National Express Group PLC (until 25 May 2011)	c2c
Merseyrail Electrics	Arriva Trains Ltd (until 20 July 2003)	Arriva Trains Merseyside
Midland Main Line	National Express Group plc (until 27 April 2008)	Midland Mainline

North London Railways	National Express Group PLC (until 1 September 2004)	Silverlink Train Services
North West Regional Railways	First Group plc (until 1 April 2004)	First North Western
Regional Railways North East	Arriva Trains Ltd (until 18 February 2003)	Arriva Trains Northern
Scotrail	National Express Group PLC (until 30 March 2004)	ScotRail
South Central	GoVia Ltd (Go-Ahead/Keolis). (until May 2010)	South Central
South Eastern	Connex Transport UK Ltd. (until 12 October 2011)	Connex
South West	Stagecoach Holdings PLC (until 3 February 2007)	South West Trains
Thames	Go-Ahead Group (until 12 April 2004)	Thames Trains
Thameslink	GoVia Ltd. (until 1 April 2004)	Thameslink Rail
Wales & Borders§	National Express Group PLC (until 30 April 2004)	Wales & Borders Trains
Wessex Trains§	National Express Group PLC (until 30 April 2004)	Wessex Trains
West Anglia Great Northern	National Express Group PLC (until 4 April 2004)	WAGN

Notes: * Franchise to be renegotiated by April 2004.
§ For the present, Wales & West PassengerTrains is the legal trading name of Wessex Trains and the Cardiff Railway Company is the legal name for Wales & Borders.

A major reorganisation of franchises is under way. See **entrain** for developments.

The following operators run non-franchised services only:

Operator	Trading Name	Route
British Airports Authority	Heathrow Express	London Paddington–Heathrow Airport
Hull Trains	Hull Trains	London King's Cross–Hull
West Coast Railway Co.	West Coast Railway	Fort William–Mallaig. York–Scarborough

INTERNATIONAL PASSENGER OPERATIONS

Eurostar (UK) operates international passenger-only services between the United Kingdom and continental Europe, jointly with French National Railways (SNCF) and Belgian National Railways (SNCB/NMBS). Eurostar (UK) is a subsidiary of London & Continental Railways, which is jointly owned by National Express Group PLC and British Airways.

In addition, a service for the conveyance of accompanied road vehicles through the Channel Tunnel is provided by the tunnel operating company, Eurotunnel.

FREIGHT TRAIN OPERATIONS

The following operators operate freight train services under 'Open Access' arrangements:

English Welsh & Scottish Railway Ltd. (EWS)
Freightliner Ltd.
GB Railfreight Ltd.
Direct Rail Services Ltd.

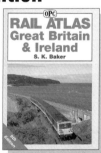

INTRODUCTION

DMU CLASSES

DMU Classes are listed in class number order. Principal details and dimensions are quoted for each class in metric and/or imperial units as considered appropriate bearing in mind common usage in the UK.

All dimensions and weights are quoted for vehicles in an 'as new' condition with all necessary supplies (e.g. oil, water, sand) on board. Dimensions are quoted in the order Length – Width. All lengths quoted are over buffers or couplers as appropriate. All width dimensions quoted are maxima.

The following abbreviations are used in class headings and also throughout this publication:

BR	British Railways.	kW	kilowatts.
BSI	Bergische Stahl Industrie.	lbf	pounds force.
DEMU	Diesel-electric multiple unit.	mm.	millimetres.
DMU	Diesel multiple unit (general term).	m.	metres.
GWR	Great Western Railway.	m.p.h.	miles per hour.
h.p.	horsepower.	r.p.m.	revolutions per minute.
Hz	Hertz.	RSL	Rolling Stock Library.
kN	kilonewtons.	t.	tonnes.
km/h	kilometres per hour.	V	volts.

NUMERICAL LISTINGS

DMUs are listed in numerical order of set – using current numbers as allocated by the RSL. Individual 'loose' vehicles are listed in numerical order after vehicles formed into fixed formations. Where numbers carried are different from those officially allocated these are noted in class headings where appropriate. Where sets or vehicles have been renumbered in recent years, former numbering detail is shown in parentheses. Each entry is laid out as in the following example:

Set No.	Detail	Livery	Owner	Operation	Depot	Formation		Name
150 257	r*	**AR**	P	*AR*	NC	52257	57257	QUEEN BOADICEA

Detail Differences. Detail differences which currently affect the areas and types of train which vehicles may work are shown, plus differences in interior layout. Where such differences occur within a class, these are shown either in the heading information or alongside the individual set or vehicle number. The following standard abbreviation is used:

r Radio Electronic Token Block (RETB) equipment.

In all cases use of the above abbreviations indicates the equipment indicated is normally operable. Meaning of non-standard abbreviations is detailed in individual class headings.

Set Formations. Regular set formations are shown where these are normally maintained. Readers should note set formations might be temporarily varied from time to time to suit maintenance and/or operational requirements. Vehicles shown as 'Spare' are not formed in any regular set formation.

Codes. Codes are used to denote the livery, owner, operation and depot of each unit. Details of these will be found in section 5 of this book. Where a unit or spare car is off-lease, the operation column will be left blank. (S) denotes stored.

Names. Only names carried with official sanction are listed. As far as possible names are shown in UPPER/lower case characters as actually shown on the name carried on the vehicle(s). Unless otherwise shown, complete units are regarded as named rather than just the individual car(s) which carry the name.

GENERAL INFORMATION

CLASSIFICATION AND NUMBERING

First generation ('Heritage') DMUs are classified in the series 100–139.
Second generation DMUs are classified in the series 140–199.
Diesel-electric multiple units are classified in the series 200–249.
Service units are classified in the series 930–999.
First and second generation individual cars are numbered in the series 50000–59999 and 79000–79999.

DEMU individual cars are numbered in the series 60000–60999, except for a few former EMU vehicles which retain their EMU numbers.

Service stock individual cars are numbered in the series 975000–975999 and 977000–977999, although this series is not exclusively used for DMU vehicles.

OPERATING CODES

These codes are used by train operating company staff to describe the various different types of vehicles and normally appear on data panels on the inner (i.e. non driving) ends of vehicles.

The first part of the code describes whether or not the car has a motor or a driving cab as follows:

DM Driving motor.
M   Motor
DT Driving trailer
T   Trailer

The next letter is a 'B' for cars with a brake compartment.

This is followed by the saloon details:

F   First
S   Standard
C   Composite
so denotes a semi-open vehicle (part compartments, part open). All other vehicles are assumed to consist solely of open saloons.

L denotes a vehicles with a lavatory compartment.

Finally vehicles with a buffet are suffixed RB or RMB for a miniature buffet.

Where two vehicles of the same type are formed within the same unit, the above codes may be suffixed by (A) and (B) to differentiate between the vehicles.

A composite is a vehicle containing both first and standard class accommodation, whilst a brake vehicle is a vehicle containing separate specific accommodation for the conductor.

Special Note: Where vehicles have been declassified, the correct operating code which describes the actual vehicle layout is quoted in this publication.

DESIGN CODES AND DIAGRAM CODES

For each type of vehicle the RSL issues a seven character 'Design Code' consisting of two letters plus four numbers and a suffix letter. (e.g. DP2010A). The first five characters of the Design Code are known as the 'Diagram Code' and these are quoted in this publication in sub-headings. The meaning of the various characters of the Design Code is as follows:

First Character

D   Diesel Multiple Unit vehicle.

Second Character

B   DEMU Driving motor passenger vehicle with brake compartment.
C   DEMU Driving motor passenger vehicle.
D   DEMU Non-driving motor passenger vehicle.
E   DEMU Driving trailer passenger vehicle.
F   DEMU Driving motor passenger vehicle (tilting).
G   DEMU Non-driving motor passenger vehicle (tilting).
H   DEMU Trailer passenger vehicle.
P   DMU (excl. DEMU) Driving motor passenger vehicle.
Q   DMU (excl. DEMU) Driving motor passenger vehicle with brake compartment.
R   DMU (excl. DEMU) Non-driving motor passenger vehicle.
S   DMU (excl. DEMU) Driving trailer passenger vehicle.
T   DMU (excl. DEMU) Trailer passenger vehicle.
X   DMU (excl. DEMU) Single unit railcar.
Z   All types of service vehicle.

Third Character

1   First class accommodation.
2   Standard class accommodation.
3   Composite accommodation.
5   No passenger accommodation.

Fourth & Fifth Characters

These distinguish between different designs of vehicle, each design being allocated a unique two digit number.

BUILD DETAILS

Lot Numbers

Vehicles ordered under the auspices of BR were allocated a Lot (batch) number when ordered and these are quoted in class headings and sub-headings.

ACCOMMODATION

The information given in class headings and sub-headings is in the form F/S nT (or TD) nW. For example 12/54 1T 1W denotes 12 first class and 54 standard class seats, 1 toilet and 1 wheelchair space. The seating layout of open saloons is shown as 2+1, 2+2 or 3+2 as the case may be. Where units have first class accommodation as well as standard and the layout is different for each class then these are shown separately prefixed by '1:' and '2:'. Compartments are always three seats a side in first class and four aside in standard class in EMUs. TD denotes a toilet suitable for a disabled person.

1. DIESEL MECHANICAL & DIESEL HYDRAULIC UNITS

1.1. FIRST GENERATION UNITS

CLASS 101   METRO-CAMMELL

First generation units still in service with First North Western.
DMBS–DMSL.
Construction: Aluminium alloy body on steel underframe.
Engines: Two Leyland 680/1 of 112 kW (150 h.p.) at 1800 r.p.m. per power car.
Transmission: Mechanical. Cardan shaft and freewheel to a four-speed epicyclic gearbox with a further cardan shaft to the final drive, each engine driving the inner axle of one bogie.
Brakes: Vacuum.
Gangways: British Standard (Midland scissors type). Within unit only.
Bogies: DD15 (motor) and DT11 (trailer).
Couplers: Screw couplings.
Dimensions: 18.49 x 2.82 m.
Seating Layout: 3+2 mainly unidirectional.
Doors: Manually-operated slam.
Multiple Working: 'Blue Square' coupling code. First generation vehicles may be coupled together to work in multiple up to a maximum of 6 motor cars or 12 cars in total in a formation. First generation vehicles may not be coupled in multiple with second generation vehicles.
Maximum Speed: 70 m.p.h.

51192/51205/51210. DMBS. Dia. DQ202. Lot No. 30467 1958–59. –/52. 32.5 t.
53164. DMBS. Dia. DQ202. Lot No. 30546 1956. –/52. 32.5 t.
53204. DMBS. Dia. DQ202. Lot No. 30259 1957. –/52. 32.5 t.
53253. DMBS. Dia. DQ202. Lot No. 30266 1957. –/52. 32.5 t.
51511. DMSL. Dia. DP317. Lot No. 30501 1959. –/58. 32.5 t.
51803. DMSL. Dia. DP210. Lot No. 30588 1959. –/72 1T. 32.5 t.
53160. DMSL. Dia. DP214. Lot No. 30253 1956. –/72 1T. 32.5 t.
53170. DMSL. Dia. DP214. Lot No. 30255 1957. –/72 1T. 32.5 t.
53266. DMSL. Dia. DP210. Lot No. 30267 1957. –/72 1T. 32.5 t.
53746. DMSL. Dia. DP210. Lot No. 30271 1957. –/72 1T. 32.5 t.

Non-standard livery: 101 692 is in Caledonian style blue with yellow/orange stripes.

101 676	**RR**	A	*NW*	LO	51205	51803
101 678	**RR**	A	*NW*	LO	51210	53746
101 680	**RR**	A	*NW*	LO	53204	51511
101 685	**G**	A	*NW*	LO	53164	53160
101 692	**O**	A	*NW*	LO	53253	53170
101 693	**S**	A	*NW*	LO	51192	53266

CLASS 121   PRESSED STEEL SUBURBAN

First generation unit being refurbished for use by Chiltern on the Aylesbury–Princes Risborough services.

Construction: Steel.
Engines: Two Leyland 1595 of 112 kW (150 h.p.) at 1800 r.p.m.
Transmission: Mechanical. Cardan shaft and freewheel to a four-speed epicyclic gearbox and final drive.
Brakes: Vacuum.
Gangways: Non gangwayed single cars with cabs at each end.
Bogies: DD10.
Couplers: Screw couplings.
Dimensions: 20.45 x 2.82 m.
Seating Layout: 3+2 facing.
Doors: Manually-operated slam.
Maximum Speed: 70 m.p.h.

55020. DMBS. Dia. DX201. Lot No. 30518 1960. –/65. 38.0 t.

| 121 020 | N | CR | *CR* | AL | 55020 |

1.2. SECOND GENERATION UNITS

All units in this section have air brakes and are equipped with public address, with transmission equipment on driving vehicles and flexible diaphragm gangways. Except where otherwise stated, transmission is Voith 211r hydraulic with a cardan shaft to a Gmeinder GM190 final drive.

CLASS 142   PACER   BREL DERBY/LEYLAND

DMS–DMSL.

Construction: Steel. Built from Leyland National bus parts on four-wheeled underframes.
Engines: One Cummins LTA10-R of 172 kW (230 h.p.) at 2100 r.p.m. (* One Perkins 2006-TWH of 172 kW (230 h.p.) at 2100 r.p.m.).
Couplers: BSI at outer ends, bar within unit.
Seating Layout: 3+2 mainly unidirectional bus style.
Dimensions: 15.66 x 2.80 m.

Gangways: Within unit only.   **Wheel Arrangement:** 1-A A-1.
Doors: Twin-leaf inward pivoting.   **Maximum Speed:** 75 m.p.h.
Multiple Working: Classes 142, 143, 144, 150, 153, 155, 156, 158, 159.

55542–55591. DMS. Dia. DP234 (s DP271). Lot No. 31003 1985–86. –/62. (s –/58, t –/53 1W, u –/54, v –/55) 23.26 t.
55592–55641. DMSL. Dia. DP235 (s DP272). Lot No. 31004 1985–86. –/59 1T. (s –/50 1T, u –/60 1T, v –/51 1T) 24.97 t.
55701–55746. DMS. Dia. DP234 (s DP271). Lot No. 31013 1986–87. –/62. (s –/58, t –/53 1W, u –/54, v–/55) 23.26 t.
55747–55792. DMSL. Dia. DP235 (s DP272). Lot No. 31014 1986–87. –/59 1T. (s –/50 1T, u –/60 1T, v–/51 iT) 24.97 t.

s Fitted with 2+2 individual high-backed seating.
t First North Western facelifed units – DMS fitted with luggage rack and wheelchair space.
u Merseytravel units – Fitted with 3+2 individual low-back seating.
v Refurbished Valley Lines units. Fitted with 2+2 individual Chapman seating.

142 001	t	NW	A	NW	NH	55542	55592	
142 002	v	VL	A	VL	CF	55543	55593	
142 003		GM	A	NW	NH	55544	55594	
142 004	t	NW	A	NW	NH	55545	55595	
142 005	t	NW	A	NW	NH	55546	55596	
142 006	v	VL	A	VL	CF	55547	55597	
142 007	t	NW	A	NW	NH	55548	55598	
142 009	t	NW	A	NW	NH	55550	55600	Newton Heath 125 1876–2001
142 010	v	VL	A	VL	CF	55551	55601	
142 011	t	NW	A	NW	NH	55552	55602	
142 012	t	NW	A	NW	NH	55553	55603	
142 013		GM	A	NW	NH	55554	55604	
142 014	t	NW	A	NW	NH	55555	55605	
142 015	s	RR	A	AV	HT	55556	55606	
142 016	s	RR	A	AV	HT	55557	55607	
142 017	s	TW	A	AV	HT	55558	55608	
142 018	s	TW	A	AV	HT	55559	55609	
142 019	s	TW	A	AV	HT	55560	55610	
142 020	s	TW	A	AV	HT	55561	55611	
142 021	s	TW	A	AV	HT	55562	55612	
142 022	s	TW	A	AV	HT	55563	55613	
142 023	t	NW	A	NW	NH	55564	55614	
142 024	s	RR	A	AV	HT	55565	55615	
142 025	s	NS	A	AV	HT	55566	55616	
142 026	s	NS	A	AV	HT	55567	55617	
142 027	t	GM	A	NW	NH	55568	55618	
142 028	t	NW	A	NW	NH	55569	55619	
142 029		GM	A	NW	NH	55570	55620	
142 030	t	NW	A	NW	NH	55571	55621	
142 031	t	NW	A	NW	NH	55572	55622	
142 032	t	NW	A	NW	NH	55573	55623	
142 033	t	NW	A	NW	NH	55574	55624	
142 034	t	NW	A	NW	NH	55575	55625	
142 035	t	NW	A	NW	NH	55576	55626	
142 036	t	NW	A	NW	NH	55577	55627	
142 037	t	NW	A	NW	NH	55578	55628	
142 038	t	NW	A	NW	NH	55579	55629	
142 039	t	NW	A	NW	NH	55580	55630	
142 040	t	GM	A	NW	NH	55581	55631	
142 041	u	MY	A	NW	NH	55582	55632	
142 042	u	MY	A	NW	NH	55583	55633	
142 043	u	MY	A	NW	NH	55584	55634	
142 044	u	MY	A	NW	NH	55585	55635	
142 045	u	MY	A	NW	NH	55586	55636	

142 046	u	**MY**	A	*NW*	NH	55587 55637
142 047	u	**MY**	A	*NW*	NH	55588 55638
142 048	u	**MY**	A	*NW*	NH	55589 55639
142 049	u	**MY**	A	*NW*	NH	55590 55640
142 050	s	**NS**	A	*AV*	HT	55591 55641
142 051	u	**MT**	A	*NW*	NH	55701 55747
142 052	u	**MT**	A	*NW*	NH	55702 55748
142 053	u	**MT**	A	*NW*	NH	55703 55749
142 054	u	**MT**	A	*NW*	NH	55704 55750
142 055	u	**MT**	A	*NW*	NH	55705 55751
142 056	u	**MT**	A	*NW*	NH	55706 55752
142 057	u	**MT**	A	*NW*	NH	55707 55753
142 058	u	**MT**	A	*NW*	NH	55708 55754
142 060	t	**NW**	A	*NW*	NH	55710 55756
142 061	t	**NW**	A	*NW*	NH	55711 55757
142 062	t	**GM**	A	*NW*	NH	55712 55758
142 063	t	**GM**	A	*NW*	NH	55713 55759
142 064	t	**GM**	A	*NW*	NH	55714 55760
142 065	s	**NS**	A	*AV*	HT	55715 55761
142 066	s	**NS**	A	*AV*	HT	55716 55762
142 067		**GM**	A	*NW*	NH	55717 55763
142 068	t	**NW**	A	*NW*	NH	55718 55764
142 069	v	**VL**	A	*VL*	CF	55719 55765
142 070	t	**GM**	A	*VL*	NH	55720 55766
142 071	s	**RR**	A	*AV*	HT	55721 55767
142 072	v	**VL**	A	*VL*	CF	55722 55768
142 073		**RR**	A	*VL*	CF	55723 55769
142 074	v	**VL**	A	*VL*	CF	55724 55770
142 075	v	**VL**	A	*VL*	CF	55725 55771
142 076		**RR**	A	*AV*	HT	55726 55772
142 077	v	**VL**	A	*VL*	CF	55727 55773
142 078	s	**RR**	A	*AV*	HT	55728 55774
142 079	s	**RR**	A	*AV*	HT	55729 55775
142 080	v	**VL**	A	*VL*	CF	55730 55776
142 081		**RR**	A	*VL*	CF	55731 55777
142 082		**RR**	A	*AV*	HT	55732 55778
142 083	v	**VL**	A	*VL*	CF	55733 55779
142 084	s*	**RR**	A	*AV*	HT	55734 55780
142 085	v	**VL**	A	*VL*	CF	55735 55781
142 086	s	**RR**	A	*AV*	HT	55736 55782
142 087	s	**RR**	A	*AV*	HT	55737 55783
142 088	s	**RR**	A	*AV*	HT	55738 55784
142 089	s	**RR**	A	*AV*	HT	55739 55785
142 090	s	**RR**	A	*AV*	HT	55740 55786
142 091	s	**RR**	A	*AV*	HT	55741 55787
142 092	s	**RR**	A	*VL*	CF	55742 55788
142 093	s	**RR**	A	*AV*	HT	55743 55789
142 094	s	**RR**	A	*VL*	CF	55744 55790
142 095	s	**RR**	A	*AV*	HT	55745 55791
142 096	s	**RR**	A	*AV*	HT	55746 55792

CLASS 143   PACER   ALEXANDER/BARCLAY

DMS–DMSL. Similar design to Class 142, but bodies built by W. Alexander with Barclay underframes.

Construction: Steel. Alexander bus bodywork on four-wheeled underframes.
Engines: One Cummins LTA10-R of 172 kW (230 h.p.) at 2100 r.p.m.
Couplers: BSI at outer ends, bar couplers within unit.
Seating Layout: 3+2 mainly unidirectional bus style.
Dimensions: 15.55 x 2.70 m.
Gangways: Within unit only.   **Wheel Arrangement:** 1-A A-1.
Doors: Twin-leaf inward pivoting.   **Maximum Speed:** 75 m.p.h.
Multiple Working: Classes 142, 143, 144, 150, 153, 155, 156, 158, 159.

DMS. Dia. DP236 Lot No. 31005 Andrew Barclay 1985–86. –/62 (v –/55). 24.5 t.
DMSL. Dia. DP237 Lot No. 31006 Andrew Barclay 1985–86. –/60 1T (v –/51 1T). 25.0 t.

v Refurbished Valley Lines units. Fitted with 2+2 individual Chapman seating.

143 601	v	**VL**	RD	*VL*	CF	55642	55667	
143 602	v	**VL**	P	*VL*	CF	55651	55668	
143 603		**BI**	P	*WX*	CF	55658	55669	
143 604	v	**VL**	P	*VL*	CF	55645	55670	
143 605	v	**VL**	P	*VL*	CF	55646	55671	Crimestoppers
143 606	v	**VL**	P	*VL*	CF	55647	55672	
143 607	v	**VL**	P	*VL*	CF	55648	55673	
143 608	v	**VL**	P	*VL*	CF	55649	55674	
143 609	v	**VL**	BC	*VL*	CF	55650	55675	TOM JONES
143 610	v	**VL**	RD	*VL*	CF	55643	55676	
143 611		**BI**	P	*WX*	CF	55652	55677	
143 612		**BI**	P	*WX*	CF	55653	55678	
143 613		**BI**	P	*WX*	CF	55654	55679	
143 614	v	**VL**	RD	*VL*	CF	55655	55680	
143 615	v	**VL**	P	*VL*	CF	55656	55681	
143 616	v	**VL**	P	*VL*	CF	55657	55682	
143 617		**BI**	RI	*WX*	CF	55644	55683	
143 618		**BI**	RI	*WX*	CF	55659	55684	
143 619		**BI**	RI	*WX*	CF	55660	55685	
143 620		**BI**	P	*WX*	CF	55661	55686	
143 621		**BI**	P	*WX*	CF	55662	55687	
143 622		**BI**	P	*WX*	CF	55663	55688	
143 623		**BI**	P	*WX*	CF	55664	55689	
143 624	v	**VL**	P	*VL*	CF	55665	55690	
143 625	v	**VL**	P	*VL*	CF	55666	55691	Valley Kids

CLASS 144   PACER   ALEXANDER/BREL DERBY

DMS–DMSL or DMS–MS–DMSL. As Class 143, but underframes built by BREL.

Construction: Steel. Alexander bus bodywork on four-wheeled underframes.
Engines: One Cummins LTA10-R of 172 kW (230 h.p.) at 2100 r.p.m.

Couplers: BSI at outer ends, bar couplers within unit.
Seating Layout: 3+2 mainly unidirectional bus style.
Dimensions: 15.55 x 2.73 m.

Gangways: Within unit only.	**Wheel Arrangement**: 1-A A-1.
Doors: Twin-leaf inward pivoting.	**Maximum Speed**: 75 m.p.h.

Multiple Working: Classes 142, 143, 144, 150, 153, 155, 156, 158, 159.

DMS. Dia. DP240 Lot No. 31015 BREL Derby 1986–87. –/62 1W (s –48 1W) 24.2 t.
MS. Dia. DR205 Lot No. BREL Derby 31037 1987. –/73 22.6 t.
DMSL. Dia. DP241 Lot No. BREL Derby 31016 1986–87. –/60 1T (s –45 1T) 25.0 t.

s Refurbished Arriva Trains Northern units. Fitted with 2+2 individual Chapman seating.

Note: The centre cars of the 3-car units are owned by West Yorkshire PTE, although managed by Porterbrook Leasing Company.

144 001	s	**YP**	P	*AV*	NL	55801		55824
144 002		**WY**	P	*AV*	NL	55802		55825
144 003		**WY**	P	*AV*	NL	55803		55826
144 004		**WY**	P	*AV*	NL	55804		55827
144 005		**WY**	P	*AV*	NL	55805		55828
144 006	s	**YP**	P	*AV*	NL	55806		55829
144 007		**WY**	P	*AV*	NL	55807		55830
144 008		**WY**	P	*AV*	NL	55808		55831
144 009		**WY**	P	*AV*	NL	55809		55832
144 010	s	**YP**	P	*AV*	NL	55810		55833
144 011	s	**YP**	P	*AV*	NL	55811		55834
144 012		**RR**	P	*AV*	NL	55812		55835
144 013	s	**YP**	P	*AV*	NL	55813		55836
144 014		**WY**	P	*AV*	NL	55814	55850	55837
144 015		**WY**	P	*AV*	NL	55815	55851	55838
144 016		**WY**	P	*AV*	NL	55816	55852	55839
144 017		**WY**	P	*AV*	NL	55817	55853	55840
144 018		**WY**	P	*AV*	NL	55818	55854	55841
144 019		**WY**	P	*AV*	NL	55819	55855	55842
144 020		**WY**	P	*AV*	NL	55820	55856	55843
144 021		**WY**	P	*AV*	NL	55821	55857	55844
144 022		**WY**	P	*AV*	NL	55822	55858	55845
144 023		**WY**	P	*AV*	NL	55823	55859	55846

CLASS 150/0   SPRINTER   BREL YORK

DMSL–MS–DMS. Prototype Sprinter.

Construction: Steel.
Engines: One Cummins NT-855-R4 of 213 kW (285 h.p.) at 2100 r.p.m.
Bogies: BX8P (powered), BX8T (non-powered).
Couplers: BSI at outer ends of driving vehicles, bar non-driving ends.
Seating Layout: 3+2 (mainly unidirectional).
Dimensions: 20.06/20.18 x 2.82 m.

Gangways: Within unit only.	**Wheel Arrangement**: 2-B – 2-B – B-2.
Doors: Single-leaf sliding.	**Maximum Speed**: 75 m.p.h.

Multiple Working: Classes 142, 143, 144, 150, 153, 155, 156, 158, 159, 170.

DMSL. Dia. DP230. Lot No. 30984 1984. –/72 1T. 35.8 t.
MS. Dia. DR202. Lot No. 30986 1984. –/92. 34.4 t.
DMS. Dia. DP231. Lot No. 30985 1984. –/76. 35.6 t.

150 001	r	**CO**	A	*CT*	TS	55200	55400	55300
150 002	r	**CO**	A	*CT*	TS	55201	55401	55301

CLASS 150/1   SPRINTER   BREL YORK

DMSL–DMS or DMSL–DMSL–DMS or DMSL–DMS–DMS.

Construction: Steel.
Engines: One Cummins NT855R5 of 213 kW (285 h.p.) at 2100 r.p.m.
Bogies: BP38 (powered), BT38 (non-powered).
Couplers: BSI.
Seating Layout: 3+2 facing as built but 150 010–150 132 were reseated with mainly unidirectional seating.
Dimensions: 19.74 x 2.82 m.
Gangways: Within unit only.
Doors: Single-leaf sliding.
Multiple Working: Classes 142, 143, 144, 150, 153, 155, 156, 158, 159, 170.

Wheel Arrangement: 2-B (– 2–B) – B-2.
Maximum Speed: 75 m.p.h.

DMSL. Dia. DP238. Lot No. 31011 1985–86. –/72 1T (s –/59 1TD, t –/71 1W 1T, u –/71 1T). 36.5 t.
DMS. Dia. DP239. Lot No. 31012 1985–86. –/76 (s –/65). 38.45 t.

Non-standard livery: 150 134 is in plain dark blue.

Notes: The centre cars of three-car units are Class 150/2 vehicles. For details see Class 150/2.
Units reliveried in **NW** livery have been refurbished with new Chapman seating.

150 010	ru	**CO**	A	*CT*	TS	52110	57226	57110
150 011	ru	**CO**	A	*CT*	TS	52111	52204	57111
150 012	ru	**CO**	A	*CT*	TS	52112	57206	57112
150 013	ru	**CO**	A	*CT*	TS	52113	57226	57113
150 014	ru	**CO**	A	*CT*	TS	52114	57204	57114
150 015	ru	**CO**	A	*CT*	TS	52115	52206	57115
150 016	ru	**CO**	A	*CT*	TS	52116	57212	57116
150 017	ru	**CO**	A	*CT*	TS	52117	57109	57117
150 018	ru	**CO**	A	*CT*	TS	52118	52220	57118
150 019	ru	**CO**	A	*CT*	TS	52119	52220	57119
150 101	ru	**CO**	A	*CT*	TS	52101	57101	
150 102	ru	**CO**	A	*CT*	TS	52102	57102	
150 103	ru	**CO**	A	*CT*	TS	52103	57103	
150 104	ru	**CO**	A	*CT*	TS	52104	57104	
150 105	ru	**CO**	A	*CT*	TS	52105	57105	
150 106	r	**CO**	A	*CT*	TS	52106	57106	
150 107	r	**CO**	A	*CT*	TS	52107	57107	
150 108	ru	**CO**	A	*CT*	TS	52108	57108	
150 109	ru	**CO**	A	*CT*	TS	52109	57109	
150 120	t	**SL**	A	*SL*	BY	52120	57120	

▲ "Heritage" Class 101s are still at work in the Manchester area for First North Western. On 02/07/02 101 676 arrives at Manchester Piccadilly with the 13.52 from Rose Hill. **Chris Booth**

▼ Several 142s now sport the Valley Lines livery. On 10/09/02 142 085 pauses at Taffs Well with a Barry Island service. **Bob Sweet**

All Wessex Trains Class 143s are now in the deep blue "Visit Bristol" livery. On 25/04/02 143 620 approaches Lawrence Hill with the 10.30 Bristol Temple Meads–Avonmouth.

Hugh Ballantyne

▲ All Class 144s are operated by Arriva Trains Northern and are undergoing refurbishment at Neville Hill during which sets are being repainted into a new West Yorkshire PTE livery. The first unit so painted, 144 006 is seen arriving at Meadowhall with the 10.08 Bridlington–Sheffield on 20/06/02. **Peter Fox**

▼ ScotRail liveried 150 262 is seen at Haymarket with a service to Edinburgh Waverley from Bathgate on 21/05/01. **Jason Rogers**

North Western Trains / First North Western liveried 150 150 is seen with a Class 156 approaching Manchester Oxford Road with a Buxton–Blackpool North service on 08/09/02. **Jason Rogers**

▲ Anglia's Class 153s are now being painted into their corporate livery. On 04/09/02 153 322 is seen near Whittlesea with a Peterborough–Ipswich service.
John Rudd

▼ 155 345 passes Colton Bridge with the 10.23 York–Manchester Victoria on 28/08/01.
John Teasdale

▲ Carmine and Cream liveried 156 495 arrives at Glasgow Central on 14/05/02.
Gavin Morrison

▼ All First North Western 156s still carry the old Regional Railways North West 'green stripe' livery, albeit with FNW branding. On 14/09/02 156 452 is seen leaving Workington with the 09.15 Whitehaven–Carlisle. **Robert Pritchard**

ScotRail liveried 158 728 passes Greenhill Lower Junction with the 18.19 Stirling–Glasgow Queen Street on 05/08/02.
Paul Robertson

▲ Wessex Trains are painting their 158s into a silver livery with purple doors (Wales & Borders' 158s are in a similar livery with blue doors). On 02/08/02 158 869 arrives at Bath with the 14.24 Portsmouth–Cardiff. **Hugh Ballantyne**

▼ South West Trains liveried 159 016 + 159 010 leave Newton Abbot on 06/07/02 with the 10.00 Brighton–Paignton. **Paul Robertson**

150 121	ru	CO	A	CT	TS	52121	57121	
150 122	ru	CO	A	CT	TS	52122	57122	
150 123	t	SL	A	SL	BY	52123	57123	
150 124	ru	CO	A	CT	TS	52124	57124	
150 125	ru	CO	A	CT	TS	52125	57125	
150 126	ru	CO	A	CT	TS	52126	57126	
150 127	t	SL	A	SL	BY	52127	57127	Bletchley TMD
150 128	t	SL	A	SL	BY	52128	57128	Community Forest
150 129	t	SL	A	SL	BY	52129	57129	MARSTON VALE
150 130	t	SL	A	SL	BY	52130	57130	Bedford–Bletchley 150
150 131	t	SL	A	SL	BY	52131	57131	LESLIE CRABBE
150 132	r	CO	A	CT	TS	52132	57132	
150 133	s	NW	A	NW	NH	52133	57133	
150 134	s	O	A	NW	NH	52134	57134	
150 135	s	NW	A	NW	NH	52135	57135	
150 136	s	NW	A	NW	NH	52136	57136	
150 137	s	NW	A	NW	NH	52137	57137	
150 138	s	NW	A	NW	NH	52138	57138	
150 139	s	NW	A	NW	NH	52139	57139	
150 140	s	NW	A	NW	NH	52140	57140	
150 141	s	NW	A	NW	NH	52141	57141	
150 142	s	NW	A	NW	NH	52142	57142	
150 143	s	NW	A	NW	NH	52143	57143	
150 144	s	NW	A	NW	NH	52144	57144	
150 145	s	NW	A	NW	NH	52145	57145	
150 146	s	NW	A	NW	NH	52146	57146	
150 147	s	NW	A	NW	NH	52147	57147	
150 148	s	NW	A	NW	NH	52148	57148	
150 149	s	NW	A	NW	NH	52149	57149	
150 150	s	NW	A	NW	NH	52150	57150	

CLASS 150/2   SPRINTER   BREL YORK

DMSL–DMS.

Construction: Steel.
Engines: One Cummins NT855R5 of 213 kW (285 h.p.) at 2100 r.p.m.
Bogies: BP38 (powered), BT38 (non-powered).
Couplers: BSI.
Seating Layout: 3+2 (2+2 v)mainly unidirectional seating.
Dimensions: 19.74 x 2.82 m.
Gangways: Throughout.   **Wheel Arrangement:** 2-B – B-2.
Doors: Single-leaf sliding.   **Maximum Speed:** 75 m.p.h.
Multiple Working: Classes 142, 143, 144, 150, 153, 155, 156, 158, 159, 170.

DMSL. Dia. DP242. Lot No. 31017 1986–87. –/73 1T (s –/70 1TD, v –/60 1T). 35.8 t.
DMS. Dia. DP243. Lot No. 31018 1986–87. –/76 (* –/68, s –/62, v –/56). 34.9 t.

Units in **NW** livery have been refurbished with new Chapman seating.
v Refurbished Valley Lines/Wessex Trains units with 2+2 Chapman seating.

150 201	s	NW	A	NW	NH	52201	57201
150 202		CO	A	CT	TS	52202	57202

150 203	s	**NW**	A	*NW*	NH	52203	57203	
150 205	s	**NW**	A	*NW*	NH	52205	57205	
150 207	s	**NW**	A	*NW*	NH	52207	57207	
150 208		**SR**	P	*SR*	HA	52208	57208	
150 210		**CO**	A	*CT*	TS	52210	57210	
150 211	s	**NW**	A	*NW*	NH	52211	57211	
150 213	r*	**AR**	P	*AR*	NC	52213	57213	LORD NELSON
150 214		**CO**	A	*CT*	TS	52214	57214	
150 215	s	**NW**	A	*NW*	NH	52215	57215	
150 216		**CO**	A	*CT*	TS	52216	57216	
150 217	r*	**AR**	P	*AR*	NC	52217	57217	OLIVER CROMWELL
150 218	s	**NW**	A	*NW*	NH	52218	57218	
150 219	r	**RR**	P	*WX*	CF	52219	57219	
150 221	rv	**WZ**	P	*WX*	CF	52221	57221	
150 222	s	**NW**	A	*NW*	NH	52222	57222	
150 223	s	**NW**	A	*WX*	CF	52223	57223	
150 224	s	**NW**	A	*NW*	NH	52224	57224	
150 225	s	**NW**	A	*NW*	NH	52225	57225	
150 227	r*	**AR**	P	*AR*	NC	52227	57227	SIR ALF RAMSEY
150 228		**RR**	P	*AV*	NL	52228	57228	
150 229	r*	**AR**	P	*AR*	NC	52229	57229	GEORGE BORROW
150 230	r	**RR**	P	*WX*	CF	52230	57230	
150 231	r*	**PS**	P	*AR*	NC	52231	57231	KING EDMUND
150 232	r	**RR**	P	*WX*	CF	52232	57232	
150 233	r	**RR**	P	*WX*	CF	52233	57233	
150 234	r	**RR**	P	*WX*	CF	52234	57234	
150 235	r*	**PS**	P	*AR*	NC	52235	57235	CARDINAL WOLSEY
150 236	r	**RR**	P	*WX*	CF	52236	57236	
150 237	r*	**AR**	P	*AR*	NC	52237	57237	HEREWARD THE WAKE
150 238	r	**RR**	P	*WX*	CF	52238	57238	
150 239	rv	**WZ**	P	*WX*	CF	52239	57239	
150 240	r	**RR**	P	*WX*	CF	52240	57240	
150 241	rv	**WZ**	P	*WX*	CF	52241	57241	The Tarka Belle
150 242	rv	**WZ**	P	*WX*	CF	52242	57242	
150 243	r	**RR**	P	*WX*	CF	52243	57243	
150 244	rv	**WZ**	P	*WX*	CF	52244	57244	The West Cornwall Experience
150 245	r	**AR**	P	*AR*	NC	52245	57245	
150 246	r	**RR**	P	*WX*	CF	52246	57246	
150 247	r	**RR**	P	*WX*	CF	52247	57247	
150 248	r	**RR**	P	*WX*	CF	52248	57248	
150 249	r	**RR**	P	*WX*	CF	52249	57249	
150 250		**SR**	P	*SR*	HA	52250	57250	
150 251	rv	**WZ**	P	*WX*	CF	52251	57251	
150 252		**SR**	P	*SR*	HA	52252	57252	
150 253	r	**RR**	P	*WX*	CF	52253	57253	
150 254	r	**RR**	P	*WX*	CF	52254	57254	
150 255	r*	**AR**	P	*AR*	NC	52255	57255	HENRY BLOGG
150 256		**SR**	P	*SR*	HA	52256	57256	
150 257	r*	**AR**	P	*AR*	NC	52257	57257	QUEEN BOADICEA
150 258		**SR**	P	*SR*	HA	52258	57258	

150 259		SR	P	SR	HA	52259	57259	
150 260		SR	P	SR	HA	52260	57260	
150 261	r	RR	P	WX	CF	52261	57261	
150 262		SR	P	SR	HA	52262	57262	
150 263	r	RR	P	WX	CF	52263	57263	
150 264		SR	P	SR	HA	52264	57264	
150 265	r	RR	P	WX	CF	52265	57265	
150 266	r	RR	P	WX	CF	52266	57266	
150 267	rv	VW	P	VL	CF	52267	57267	
150 268		RR	P	AV	NL	52268	57268	
150 269		RR	P	AV	NL	52269	57269	
150 270		RR	P	AV	NL	52270	57270	
150 271		RR	P	AV	NL	52271	57271	
150 272		RR	P	AV	NL	52272	57272	
150 273		RR	P	AV	NL	52273	57273	
150 274		RR	P	AV	NL	52274	57274	
150 275		RR	P	AV	NL	52275	57275	
150 276		RR	P	AV	NL	52276	57276	
150 277		RR	P	AV	NL	52277	57277	
150 278	rv	VW	P	VL	CF	52278	57278	The Millennium Stadium
150 279	v	VW	P	VL	CF	52279	57279	Rhondda Heritage Park
150 280	v	VW	P	VL	CF	52280	57280	University of Glamorgan/ Prifysgol Morgannwg
150 281	v	VW	P	VL	CF	52281	57281	Cardiff Castle
150 282	v	VW	P	VL	CF	52282	57282	Caerphilly Castle
150 283		SR	P	SR	HA	52283	57283	
150 284		SR	P	SR	HA	52284	57284	
150 285		SR	P	SR	HA	52285	57285	

CLASS 153　SUPER SPRINTER　LEYLAND BUS

DMSL. Converted by Hunslet-Barclay, Kilmarnock from Class 155 two-car units.

Construction: Steel. Built from Leyland National bus parts on bogied underframes.
Engine: One Cummins NT855R5 of 213 kW (285 h.p.) at 2100 r.p.m.
Bogies: One P3-10 (powered) and one BT38 (non-powered).
Couplers: BSI.
Seating Layout: 2+2 facing/unidirectional.
Dimensions: 23.21 x 2.70 m.
Gangways: Throughout.　　　　　　　　**Wheel Arrangement:** 2-B.
Doors: Single-leaf sliding plug.　　　　**Maximum Speed:** 75 m.p.h.
Multiple Working: Classes 142, 143, 144, 150, 153, 155, 156, 158, 159, 170.

52301–52335. DMSL. Dia. DX203. Lot No. 31026 1987–88. Converted under Lot No. 31115 1991–2. –/75 1T 1W (* –/66 1T 1W). 41.2 t.
57301–57335. DMSL. Dia. DX203. Lot No. 31027 1987–88. Converted under Lot No. 31115 1991–2. –/75 1T (* –/66 1T). 41.2 t.

Advertising Livery: 153 314 Norfolk and Norwich Festival (Black and orange).

Notes:

Cars numbered in the 573XX series were renumbered by adding 50 to their original number so that the last two digits correspond with the set number. Central Trains and First North Western units have been fitted with new Chapman seating. Wales & Borders/Wessex Trains units have been reseated with seats removed from that company's Class 158 units. Arriva units have been fitted with new Richmond seating.

153 301		**AV**	A	*AV*	NL	52301	
153 302	r	**DC**	A	*WX*	CF	52302	
153 303	r	**HW**	A	*WB*	CF	52303	
153 304		**AV**	A	*AV*	NL	52304	
153 305	r	**WX**	A	*WX*	CF	52305	
153 306	r*	**PS**	P	*AR*	NC	52306	EDITH CAVELL
153 307		**AV**	A	*AV*	NL	52307	
153 308	r	**DC**	A	*WX*	CF	52308	
153 309	r*	**PS**	P	*AR*	NC	52309	GERARD FIENNES
153 310	r	**NW**	P	*NW*	NH	52310	
153 311	r*	**PS**	P	*AR*	NC	52311	JOHN CONSTABLE
153 312	r	**HW**	A	*WB*	CF	52312	
153 313		**NW**	P	*NW*	NH	52313	
153 314	r*	**AL**	P	*AR*	NC	52314	DELIA SMITH
153 315		**AV**	A	*AV*	NL	52315	
153 316		**NW**	P	*NW*	NH	52316	
153 317		**AV**	A	*AV*	NL	52317	
153 318	r	**WX**	A	*WX*	CF	52318	
153 319		**AV**	A	*AV*	NL	52319	
153 320	r	**HW**	P	*WB*	CF	52320	
153 321	r	**HW**	P	*WB*	CF	52321	
153 322	r*	**AR**	P	*AR*	NC	52322	BENJAMIN BRITTEN
153 323	r	**HW**	P	*WB*	CF	52323	
153 324		**NW**	P	*NW*	NH	52324	
153 325	r	**RR**	P	*CT*	TS	52325	
153 326	r*	**PS**	P	*AR*	NC	52326	TED ELLIS
153 327	r	**HW**	A	*WB*	CF	52327	
153 328		**AV**	A	*AV*	NL	52328	
153 329	r	**RR**	P	*CT*	TS	52329	
153 330		**NW**	P	*NW*	NH	52330	
153 331		**AV**	A	*AV*	NL	52331	
153 332		**NW**	P	*NW*	NH	52332	
153 333	r	**RR**	P	*CT*	TS	52333	
153 334	r	**RR**	P	*CT*	TS	52334	
153 335	r*	**PS**	P	*AR*	NC	52335	MICHAEL PALIN
153 351		**AV**	A	*AV*	NL	57351	
153 352		**AV**	A	*AV*	NL	57352	
153 353	r	**HW**	A	*WB*	CF	57353	
153 354	r	**RR**	P	*CT*	TS	57354	
153 355	r	**WX**	A	*WX*	CF	57355	
153 356	r	**RR**	P	*CT*	TS	57356	
153 357		**AV**	A	*AV*	NL	57357	
153 358		**NW**	P	*NW*	NH	57358	

153 359		NW	P	NW	NH	57359
153 360		NW	P	NW	NH	57360
153 361		NW	P	NW	NH	57361
153 362	r	HW	A	WB	CF	57362
153 363		NW	P	NW	NH	57363
153 364	r	RR	P	CT	TS	57364
153 365	r	RR	P	CT	TS	57365
153 366	r	RR	P	CT	TS	57366
153 367		NW	P	NW	NH	57367
153 368	r	WX	A	WX	CF	57368
153 369	r	RR	P	CT	TS	57369
153 370	r	WX	A	WX	CF	57370
153 371	r	RR	P	CT	TS	57371
153 372	r	WX	A	WX	CF	57372
153 373	r	WX	A	WX	CF	57373
153 374	r	DC	A	WX	CF	57374
153 375	r	RR	P	CT	TS	57375
153 376	r	RR	P	CT	TS	57376
153 377	r	DC	A	WX	CF	57377
153 378		AV	A	AV	NL	57378
153 379	r	RR	P	CT	TS	57379
153 380	r	DC	A	WX	CF	57380
153 381	r	RR	P	CT	TS	57381
153 382	r	DC	A	WX	CF	57382
153 383	r	RR	P	CT	TS	57383
153 384	r	RR	P	CT	TS	57384
153 385	r	RR	P	CT	TS	57385

CLASS 155   SUPER SPRINTER   LEYLAND BUS

DMSL–DMS.

Construction: Steel. Built from Leyland National bus parts on bogied underframes.
Engines: One Cummins NT855R5 of 213 kW (285 h.p.) at 2100 r.p.m.
Bogies: One P3-10 (powered) and one BT38 (non-powered).
Couplers: BSI.
Seating Layout: 2+2 facing/unidirectional.
Dimensions: 23.21 x 2.70 m.
Gangways: Throughout.   **Wheel Arrangement:** 2-B – B-2.
Doors: Single-leaf sliding plug.   **Maximum Speed:** 75 m.p.h.
Multiple Working: Classes 142, 143, 144, 150, 153, 155, 156, 158, 159, 170.

DMSL. Dia. DP248. Lot No. 31057 1988. –/80 1TD 1W. 39.0 t.
DMS. Dia. DP249. Lot No. 31058 1988. –/80. 38.7 t.

Note: These units are owned by West Yorkshire PTE, although managed by Porterbrook Leasing Company.

155 341	WY	P	AV	NL	52341	57341
155 342	WY	P	AV	NL	52342	57342
155 343	WY	P	AV	NL	52343	57343
155 344	WY	P	AV	NL	52344	57344

155 345	**WY**	P	*AV*	NL	52345	57345
155 346	**WY**	P	*AV*	NL	52346	57346
155 347	**WY**	P	*AV*	NL	52347	57347

CLASS 156 SUPER SPRINTER METRO-CAMMELL

DMSL–DMS.

Construction: Steel.
Engines: One Cummins NT855R5 of 213 kW (285 h.p.) at 2100 r.p.m.
Bogies: One P3-10 (powered) and one BT38 (non-powered).
Couplers: BSI.
Seating Layout: 2+2 facing/unidirectional.
Dimensions: 23.03 x 2.73 m.
Gangways: Throughout.   **Wheel Arrangement:** 2-B – B-2.
Doors: Single-leaf sliding plug.   **Maximum Speed:** 75 m.p.h.
Multiple Working: Classes 142, 143, 144, 150, 153, 155, 156, 158, 159, 170.

DMSL. Dia. DP244 (DP261 q). Lot No. 31028 1988–89. –/74 (t* –/72, st –/70, u –/68) 1TD 1W. 36.1 t.
DMS. Dia. DP245 (DP262 q). Lot No. 31029 1987–89. –/76 (q –/78, † –/74, tuš –/72) 35.5 t.

Notes: 156 500–514 are owned by Strathclyde PTE, although managed by Angel Train Contracts.
Central Trains and First North Western units have been fitted with new Chapman seating. Arriva units have been fitted with new Richmond seating.

156 401	r*	**RE**	P	*CT*	TS	52401	57401
156 402	r*	**RE**	P	*CT*	TS	52402	57402
156 403	r*	**RE**	P	*CT*	TS	52403	57403
156 404	r*	**RE**	P	*CT*	TS	52404	57404
156 405	r*	**RE**	P	*CT*	TS	52405	57405
156 406	r*	**RE**	P	*CT*	TS	52406	57406
156 407	r*	**CT**	P	*CT*	TS	52407	57407
156 408	r*	**RE**	P	*CT*	TS	52408	57408
156 409	r*	**RE**	P	*CT*	TS	52409	57409
156 410	r*	**RE**	P	*CT*	TS	52410	57410
156 411	r*	**RE**	P	*CT*	TS	52411	57411
156 412	r*	**RE**	P	*CT*	TS	52412	57412
156 413	r*	**RE**	P	*CT*	TS	52413	57413
156 414	r*	**RE**	P	*CT*	TS	52414	57414
156 415	r*	**RE**	P	*CT*	TS	52415	57415
156 416	r*	**RE**	P	*CT*	TS	52416	57416
156 417	r*	**RE**	P	*CT*	TS	52417	57417
156 418	r*	**RE**	P	*CT*	TS	52418	57418
156 419	r*	**RE**	P	*CT*	TS	52419	57419
156 420	s	**RN**	P	*NW*	NH	52420	57420
156 421	s	**RN**	P	*NW*	NH	52421	57421
156 422	r*	**RE**	P	*CT*	TS	52422	57422
156 423	s	**RN**	P	*NW*	NH	52423	57423
156 424	s	**RN**	P	*NW*	NH	52424	57424
156 425	s	**RN**	P	*NW*	NH	52425	57425

156 426	s	**RN**	P	*NW*	NH	52426 57426
156 427	s	**RN**	P	*NW*	NH	52427 57427
156 428	s	**RN**	P	*NW*	NH	52428 57428
156 429	s	**RN**	P	*NW*	NH	52429 57429
156 430	t	**SC**	A	*SR*	CK	52430 57430
156 431	t	**SC**	A	*SR*	CK	52431 57431
156 432	t	**SC**	A	*SR*	CK	52432 57432
156 433	t	**SC**	A	*SR*	CK	52433 57433
156 434	t	**SC**	A	*SR*	CK	52434 57434
156 435	t	**SC**	A	*SR*	CK	52435 57435
156 436	†	**SC**	A	*SR*	CK	52436 57436
156 437	rt	**SC**	A	*SR*	CK	52437 57437
156 438		**NS**	A	*AV*	NL	52438 57438
156 439	rt	**SC**	A	*SR*	CK	52439 57439
156 440	s	**RN**	P	*NW*	NH	52440 57440
156 441	s	**RN**	P	*NW*	NH	52441 57441
156 442	rt	**SC**	A	*SR*	CK	52442 57442
156 443	q	**NS**	A	*AV*	HT	52443 57443
156 444	q	**NS**	A	*AV*	HT	52444 57444
156 445	u	**SC**	A	*SR*	CK	52445 57445
156 446	rt	**SR**	A	*SR*	CK	52446 57446
156 447	ru	**SR**	A	*SR*	CK	52447 57447
156 448	q	**NS**	A	*AV*	HT	52448 57448
156 449	ru	**SR**	A	*SR*	CK	52449 57449
156 450	t	**SR**	A	*SR*	CK	52450 57450
156 451	q	**NS**	A	*AV*	HT	52451 57451
156 452	s	**RN**	P	*NW*	NH	52452 57452
156 453	ru	**SR**	A	*SR*	CK	52453 57453
156 454	q	**NS**	A	*AV*	HT	52454 57454
156 455	s	**RN**	P	*NW*	NH	52455 57455
156 456	rt	**SR**	A	*SR*	CK	52456 57456
156 457	rt	**SR**	A	*SR*	CK	52457 57457
156 458	t	**SR**	A	*SR*	CK	52458 57458
156 459	s	**RN**	P	*NW*	NH	52459 57459
156 460	s	**RN**	P	*NW*	NH	52460 57460
156 461	s	**RN**	P	*NW*	NH	52461 57461
156 462	r	**SR**	A	*SR*	CK	52462 57462
156 463	q	**NS**	A	*AV*	HT	52463 57463
156 464	s	**RN**	P	*NW*	NH	52464 57464
156 465	u	**SR**	A	*SR*	CK	52465 57465
156 466	s	**RN**	P	*NW*	NH	52466 57466
156 467	r	**SR**	A	*SR*	CK	52467 57467
156 468	q	**NS**	A	*AV*	NL	52468 57468
156 469	q	**NS**	A	*AV*	HT	52469 57469
156 470	q	**NS**	A	*AV*	NL	52470 57470
156 471	q	**NS**	A	*AV*	NL	52471 57471
156 472	q	**NS**	A	*AV*	NL	52472 57472
156 473	q	**NS**	A	*AV*	NL	52473 57473
156 474	rt	**SR**	A	*SR*	CK	52474 57474
156 475	q	**NS**	A	*AV*	NL	52475 57475
156 476	rt	**SR**	A	*SR*	CK	52476 57476

The Kilmarnock Edition

156 477	rt	**SR**	A	*SR*	CK	52477	57477
156 478	t	**SR**	A	*SR*	CK	52478	57478
156 479	q	**NS**	A	*AV*	NL	52479	57479
156 480	q	**NS**	A	*AV*	NL	52480	57480
156 481	q	**NS**	A	*AV*	NL	52481	57481
156 482	q	**NS**	A	*AV*	NL	52482	57482
156 483	q	**NS**	A	*AV*	NL	52483	57483
156 484	q	**NS**	A	*AV*	NL	52484	57484
156 485	ru	**SR**	A	*SR*	CK	52485	57485
156 486	q	**NS**	A	*AV*	NL	52486	57486
156 487	q	**NS**	A	*AV*	NL	52487	57487
156 488	q	**NS**	A	*AV*	NL	52488	57488
156 489	q	**NS**	A	*AV*	NL	52489	57489
156 490	q	**NS**	A	*AV*	NL	52490	57490
156 491	q	**NS**	A	*AV*	NL	52491	57491
156 492	rt	**SR**	A	*SR*	CK	52492	57492
156 493	rt	**SR**	A	*SR*	CK	52493	57493
156 494	§	**SC**	A	*SR*	CK	52494	57494
156 495	ru	**SC**	A	*SR*	CK	52495	57495
156 496	ru	**SR**	A	*SR*	CK	52496	57496
156 497	q	**NS**	A	*AV*	NL	52497	57497
156 498	q	**NS**	A	*AV*	NL	52498	57498
156 499	rt	**SR**	A	*SR*	CK	52499	57499
156 500	u	**SC**	A	*SR*	CK	52500	57500
156 501		**SC**	A	*SR*	CK	52501	57501
156 502		**SC**	A	*SR*	CK	52502	57502
156 503		**SC**	A	*SR*	CK	52503	57503
156 504		**SC**	A	*SR*	CK	52504	57504
156 505		**SC**	A	*SR*	CK	52505	57505
156 506		**SC**	A	*SR*	CK	52506	57506
156 507		**SC**	A	*SR*	CK	52507	57507
156 508		**SC**	A	*SR*	CK	52508	57508
156 509		**SC**	A	*SR*	CK	52509	57509
156 510		**SC**	A	*SR*	CK	52510	57510
156 511		**SC**	A	*SR*	CK	52511	57511
156 512		**SC**	A	*SR*	CK	52512	57512
156 513		**SC**	A	*SR*	CK	52513	57513
156 514		**SC**	A	*SR*	CK	52514	57514

CLASS 158/0   BREL

DMSL (B)–DMSL (A) or DMCL–DMSL or DMCL–MSL–DMSL.

Construction: Welded aluminium.
Engines: 158 701–158 814: One Cummins NTA855R of 260 kW (350 h.p.) at 1900 r.p.m.
158 863–158 872: One Cummins NTA855R of 300 kW (400 h.p.) at 2100 r.p.m.
158 815–158 862: One Perkins 2006-TWH of 260 kW (350 h.p.) at 1900 r.p.m. car.
Bogies: One BREL P4 (powered) and one BREL T4 (non-powered) per car.
Couplers: BSI.

Seating Layout: 2+2 facing/unidirectional in standard class and in ScotRail first class. 2+2 facing in Arriva Trains Northern first class, 2+1 facing/unidirectional in Virgin Cross-Country first class.
Dimensions: 22.57 x 2.70 m.
Gangways: Throughout.
Doors: Twin-leaf swing plug.
Multiple Working: Classes 142, 143, 144, 150, 153, 155, 156, 158, 159, 170.

Wheel Arrangement: 2-B – B-2.
Maximum Speed: 90 m.p.h.

DMSL (B). Dia. DP252. Lot No. 31051 BREL Derby 1989–92. –/68 1TD 1W. (†–/66 1TD 1W). Public telephone and trolley space. 38.5 t.
MSL. Dia. DR207. Lot No. 31050 BREL Derby 1991. 37.1 t. –/70 2T. 37.1 t.
DMSL (A). Dia. DP251 Lot No. 31052 BREL Derby 1989–92. –/70 († –/68; § 32/32) 1T. 37.8 t.

The above details refer to the "as built" condition. The following DMSL(B) have now been converted to DMCL as follows:

52701–52744 (ScotRail/Arriva Trains Northern). Dia. DP318. 15/51 1TD 1W (*15/53 1TD 1W).
52747–52751. (Virgin Cross-Country/Arriva Trains Northern). Dia. DP323. 9/51 1TD 1W.
52760–779/781. (Arriva Trains Northern 2-car Units). Dia. DP331. 16/48 1TD 1W.
52798–814 (Arriva Trains Northern 3-car Units). Dia. DP332. 32/32 1TD 1W.

Non-standard livery: 158 867 is in a prototype Wales & West livery (silver, blue and orange).

s Arriva Transpennine and Central Trains units refurbished with new shape seat cushions. Arriva Transpennine units also fitted with table lamps in first class.
† Fitted with new Chapman seating.

158 701	*	**SR**	P	*SR*	HA	52701	57701
158 702	*	**SR**	P	*SR*	HA	52702	57702	BBC Scotland – 75 Years
158 703	*	**SR**	P	*SR*	HA	52703	57703
158 704	*	**SR**	P	*SR*	HA	52704	57704
158 705	*	**SR**	P	*SR*	HA	52705	57705
158 706	*	**SR**	P	*SR*	HA	52706	57706
158 707	*	**SR**	P	*SR*	HA	52707	57707	Far North Line 125th ANNIVERSARY
158 708	*	**SR**	P	*SR*	HA	52708	57708
158 709	*	**SR**	P	*SR*	HA	52709	57709
158 710	*	**SR**	P	*SR*	HA	52710	57710
158 711	*	**SR**	P	*SR*	HA	52711	57711
158 712	*	**SR**	P	*SR*	HA	52712	57712
158 713	*	**SR**	P	*SR*	HA	52713	57713
158 714	*	**SR**	P	*SR*	HA	52714	57714
158 715	*	**SR**	P	*SR*	HA	52715	57715	Haymarket
158 716	*	**SR**	P	*SR*	HA	52716	57716
158 717	*	**SR**	P	*SR*	HA	52717	57717
158 718	*	**SR**	P	*SR*	HA	52718	57718
158 719	*	**SR**	P	*SR*	HA	52719	57719
158 720	*	**SR**	P	*SR*	HA	52720	57720
158 721	*	**SR**	P	*SR*	HA	52721	57721
158 722	*	**SR**	P	*SR*	HA	52722	57722

158 723	*	**SR**	P	*SR*	HA	52723	57723
158 724	*	**SR**	P	*SR*	HA	52724	57724
158 725	*	**SR**	P	*SR*	HA	52725	57725
158 726	*	**SR**	P	*SR*	HA	52726	57726
158 727	*	**SR**	P	*SR*	HA	52727	57727
158 728	*	**SR**	P	*SR*	HA	52728	57728
158 729	*	**SR**	P	*SR*	HA	52729	57729
158 730	*	**SR**	P	*SR*	HA	52730	57730
158 731	*	**SR**	P	*SR*	HA	52731	57731
158 732	*	**SR**	P	*SR*	HA	52732	57732
158 733	*	**SR**	P	*SR*	HA	52733	57733
158 734	*	**SR**	P	*SR*	HA	52734	57734
158 735	*	**SR**	P	*SR*	HA	52735	57735
158 736	*	**SR**	P	*SR*	HA	52736	57736
158 737		**TX**	P	*AV*	NL	52737	57737
158 738	*	**SR**	P	*SR*	HA	52738	57738
158 739	*	**SR**	P	*SR*	HA	52739	57739
158 740	*	**SR**	P	*SR*	HA	52740	57740
158 741	*	**SR**	P	*SR*	HA	52741	57741
158 742		**TX**	P	*AV*	NL	52742	57742
158 743		**TX**	P	*AV*	NL	52743	57743
158 744		**TX**	P	*AV*	NL	52744	57744
158 745	†	**WT**	P	*WX*	CF	52745	57745
158 746	†	**WT**	P	*WX*	CF	52746	57746	Spirit of the South West
158 747		**RE**	P	*VX*	NH	52747	57747
158 748		**RE**	P	*VX*	NH	52748	57748
158 749		**RE**	P	*VX*	NH	52749	57749
158 750		**RE**	P	*AV*	NL	52750	57750
158 751		**RE**	P	*VX*	NH	52751	57751
158 752		**NW**	P	*NW*	NH	52752	57752
158 753		**NW**	P	*NW*	NH	52753	57753
158 754		**NW**	P	*NW*	NH	52754	57754
158 755		**NW**	P	*NW*	NH	52755	57755
158 756		**NW**	P	*NW*	NH	52756	57756
158 757		**NW**	P	*NW*	NH	52757	57757
158 758		**NW**	P	*NW*	NH	52758	57758
158 759		**NW**	P	*NW*	NH	52759	57759
158 760	s	**TX**	P	*AV*	NL	52760	57760
158 761	s	**TX**	P	*AV*	NL	52761	57761
158 762	s	**TX**	P	*AV*	NL	52762	57762
158 763	s	**TX**	P	*AV*	NL	52763	57763
158 764	s	**TX**	P	*AV*	NL	52764	57764
158 765	s	**TX**	P	*AV*	NL	52765	57765
158 766	s	**TX**	P	*AV*	NL	52766	57766
158 767	s	**TX**	P	*AV*	NL	52767	57767
158 768	s	**TX**	P	*AV*	NL	52768	57768
158 769	s	**TX**	P	*AV*	NL	52769	57769
158 770	s	**TX**	P	*AV*	NL	52770	57770
158 771	s	**TX**	P	*AV*	HT	52771	57771
158 772	s	**TX**	P	*AV*	NL	52772	57772
158 773	s	**TX**	P	*AV*	NL	52773	57773

158 774	s	**TX**	P	*AV*	NL	52774	57774
158 775	s	**TX**	P	*AV*	HT	52775	57775
158 776	s	**TX**	P	*AV*	HT	52776	57776
158 777	s	**TX**	P	*AV*	HT	52777	57777
158 778	s	**TX**	P	*AV*	HT	52778	57778
158 779	s	**TX**	P	*AV*	HT	52779	57779
158 780	s	**CT**	A	*CT*	TS	52780	57780
158 781	s	**TX**	P	*AV*	HT	52781	57781
158 782	s	**CT**	A	*CT*	TS	52782	57782
158 783	s	**CT**	A	*CT*	TS	52783	57783
158 784	s	**CT**	A	*CT*	TS	52784	57784
158 785	s	**CT**	A	*CT*	TS	52785	57785
158 786	s	**CT**	A	*CT*	TS	52786	57786
158 787	s	**CT**	A	*CT*	TS	52787	57787
158 788	s	**CT**	A	*CT*	TS	52788	57788
158 789	s	**CT**	A	*CT*	TS	52789	57789
158 790	s	**CT**	A	*CT*	TS	52790	57790
158 791	s	**CT**	A	*CT*	TS	52791	57791
158 792	s	**CT**	A	*CT*	TS	52792	57792
158 793	s	**CT**	A	*CT*	TS	52793	57793
158 794	s	**CT**	A	*CT*	TS	52794	57794
158 795	s	**CT**	A	*CT*	TS	52795	57795
158 796	s	**CT**	A	*CT*	TS	52796	57796
158 797	s	**CT**	A	*CT*	TS	52797	57797
158 798	s	**TX**	P	*AV*	HT	52798	58715	57798
158 799	s	**TX**	P	*AV*	HT	52799	58716	57799
158 800	s	**TX**	P	*AV*	HT	52800	58717	57800
158 801	s	**TX**	P	*AV*	HT	52801	58701	57801
158 802	s	**TX**	P	*AV*	HT	52802	58702	57802
158 803	s	**TX**	P	*AV*	HT	52803	58703	57803
158 804	s	**TX**	P	*AV*	HT	52804	58704	57804
158 805	s	**TX**	P	*AV*	HT	52805	58705	57805
158 806	s	**TX**	P	*AV*	HT	52806	58706	57806
158 807	s	**TX**	P	*AV*	HT	52807	58707	57807
158 808	s	**TX**	P	*AV*	HT	52808	58708	57808
158 809	s	**TX**	P	*AV*	HT	52809	58709	57809
158 810	s	**TX**	P	*AV*	HT	52810	58710	57810
158 811	s	**TX**	P	*AV*	HT	52811	58711	57811
158 812	s	**TX**	P	*AV*	HT	52812	58712	57812
158 813	s	**TX**	P	*AV*	HT	52813	58713	57813
158 814	s	**TX**	P	*AV*	HT	52814	58714	57814
158 815	†	**WT**	A	*WX*	CF	52815	57815
158 816	†	**WT**	A	*WX*	CF	52816	57816
158 817	†	**WB**	A	*WB*	CF	52817	57817
158 818	†	**RE**	A	*WB*	CF	52818	57818
158 819	†	**RE**	A	*WB*	CF	52819	57819
158 820	†	**RE**	A	*WB*	CF	52820	57820
158 821	†	**RE**	A	*WB*	CF	52821	57821
158 822	†	**RE**	A	*WB*	CF	52822	57822
158 823	†	**RE**	A	*WB*	CF	52823	57823
158 824	†	**RE**	A	*WB*	CF	52824	57824

158 825	†	**RE**	A	*WB*	CF	52825	57825
158 826	†	**RE**	A	*WB*	CF	52826	57826
158 827	†	**RE**	A	*WB*	CF	52827	57827
158 828	†	**RE**	A	*WB*	CF	52828	57828
158 829	†	**WB**	A	*WB*	CF	52829	57829
158 830	†	**RE**	A	*WB*	CF	52830	57830
158 831	†	**WB**	A	*WB*	CF	52831	57831
158 832	†	**RE**	A	*WB*	CF	52832	57832
158 833	†	**WB**	A	*WB*	CF	52833	57833
158 834	†	**WB**	A	*WB*	CF	52834	57834
158 835	†	**WB**	A	*WB*	CF	52835	57835
158 836	†	**RE**	A	*WB*	CF	52836	57836
158 837	†	**RE**	A	*WB*	CF	52837	57837
158 838	†	**RE**	A	*WB*	CF	52838	57838
158 839	†	**RE**	A	*WB*	CF	52839	57839
158 840	†	**RE**	A	*WB*	CF	52840	57840
158 841	†	**RE**	A	*WB*	CF	52841	57841
158 842	†	**RE**	A	*WB*	CF	52842	57842
158 843	†	**WB**	A	*WB*	CF	52843	57843
158 844	rs	**CT**	A	*WB*	TS	52844	57844
158 845	rs	**CT**	A	*WB*	TS	52845	57845
158 846	rs	**CT**	A	*WB*	TS	52846	57846
158 847	rs	**CT**	A	*WB*	TS	52847	57847
158 848	rs	**CT**	A	*WB*	TS	52848	57848
158 849	rs	**CT**	A	*WB*	TS	52849	57849
158 850	rs	**CT**	A	*WB*	TS	52850	57850
158 851	rs	**CT**	A	*WB*	TS	52851	57851
158 852	rs	**CT**	A	*WB*	TS	52852	57852
158 853	rs	**CT**	A	*WB*	TS	52853	57853
158 854	rs	**CT**	A	*WB*	TS	52854	57854
158 855	rs	**CT**	A	*WB*	TS	52855	57855
158 856	s	**CT**	A	*CT*	TS	52856	57856
158 857	s	**CT**	A	*CT*	TS	52857	57857
158 858	s	**CT**	A	*CT*	TS	52858	57858
158 859	s	**CT**	A	*CT*	TS	52859	57859
158 860	s	**CT**	A	*CT*	TS	52860	57860
158 861	s	**CT**	A	*CT*	TS	52861	57861
158 862	s	**CT**	A	*CT*	TS	52862	57862
158 863	†	**WT**	A	*WX*	CF	52863	57863
158 864	†	**WT**	A	*WX*	CF	52864	57864
158 865	†	**WT**	A	*WX*	CF	52865	57865
158 866	†	**WT**	A	*WX*	CF	52866	57866
158 867	†	**0**	A	*WX*	CF	52867	57867
158 868	†	**WT**	A	*WX*	CF	52868	57868
158 869	†	**WT**	A	*WX*	CF	52869	57869
158 870	†	**WT**	A	*WX*	CF	52870	57870
158 871	†	**WT**	A	*WX*	CF	52871	57871
158 872	†	**WT**	A	*WX*	CF	52872	57872

CLASS 158/9   BREL

DMSL–DMS. Units leased by West Yorkshire PTE. Details as for Class 158/0 except for seating layout and toilets.

DMSL.. Dia. DP252. Lot No. 31051 BREL Derby 1990–92. –/70 1TD 1W. Public telephone and trolley space. 38.1 t.
DMS. Dia. DP251. Lot No. 31052 BREL Derby 1990–92. –/72 and parcels area. 37.8 t.

Note: These units are leased by West Yorkshire PTE and are managed by Porterbrook Leasing Company.

158 901	WY	P	AV	NL	52901	57901
158 902	WY	P	AV	NL	52902	57902
158 903	WY	P	AV	NL	52903	57903
158 904	WY	P	AV	NL	52904	57904
158 905	WY	P	AV	NL	52905	57905
158 906	WY	P	AV	NL	52906	57906
158 907	WY	P	AV	NL	52907	57907
158 908	WY	P	AV	NL	52908	57908
158 909	YN	P	AV	NL	52909	57909
158 910	WY	P	AV	NL	52910	57910

CLASS 159   BREL

DMCL–MSL–DMSL. Built as Class 158. Converted before entering passenger service to Class 159 by Rosyth Dockyard.

Construction: Welded aluminium.
Engines: One Cummins NTA855R of 300 kW (400 h.p.) at 2100 r.p.m.
Bogies: One BREL P4 (powered) and one BREL T4 (non-powered) per car.
Couplers: BSI.
Seating Layout: 1: 2+1 facing, 2: 2+2 facing/unidirectional.
Dimensions: 23.21 x 2.82 m.
Gangways: Throughout.   **Wheel Arrangement:** 2-B – B-2 – B-2.
Doors: Twin-leaf swing plug.   **Maximum Speed:** 90 m.p.h.
Multiple Working: Classes 142, 143, 144, 150, 153, 155, 156, 158, 159, 170.

DMCL. Dia. DP322. Lot No. 31051 BREL Derby 1992–93. 24/28 1TD 1W. 38.5 t.
MSL. Dia. DR209. Lot No. 31050 BREL Derby 1992–93. 38 t. –/72 2T.
DMSL. Dia. DP260. Lot No. 31052 BREL Derby 1992–93. –/72 1T and parcels area. 37.8 t.

159 001	SW	P	SW	SA	52873	58718	57873	CITY OF EXETER
159 002	SW	P	SW	SA	52874	58719	57874	CITY OF SALISBURY
159 003	SW	P	SW	SA	52875	58720	57875	TEMPLECOMBE
159 004	SW	P	SW	SA	52876	58721	57876	BASINGSTOKE AND DEANE
159 005	SW	P	SW	SA	52877	58722	57877	
159 006	SW	P	SW	SA	52878	58723	57878	
159 007	SW	P	SW	SA	52879	58724	57879	
159 008	SW	P	SW	SA	52880	58725	57880	

159 009	**SW** P *SW*	SA	52881	58726	57881
159 010	**SW** P *SW*	SA	52882	58727	57882
159 011	**SW** P *SW*	SA	52883	58728	57883
159 012	**SW** P *SW*	SA	52884	58729	57884
159 013	**SW** P *SW*	SA	52885	58730	57885
159 014	**SW** P *SW*	SA	52886	58731	57886
159 015	**SW** P *SW*	SA	52887	58732	57887
159 016	**SW** P *SW*	SA	52888	58733	57888
159 017	**SW** P *SW*	SA	52889	58734	57889
159 018	**SW** P *SW*	SA	52890	58735	57890
159 019	**SW** P *SW*	SA	52891	58736	57891
159 020	**SW** P *SW*	SA	52892	58737	57892
159 021	**SW** P *SW*	SA	52893	58738	57893
159 022	**SW** P *SW*	SA	52894	58739	57894

CLASS 165/0   NETWORK TURBO   BREL

DMCL–DMS or DMCL–MS–DMS.

Construction: Welded aluminium.
Engines: One Perkins 2006-TWH of 260 kW (350 h.p.) at 1900 r.p.m.
Bogies: BREL P3-17 (powered), BREL T3-17 (non-powered).
Couplers: BSI.
Seating Layout: 1: 2+2 facing, 2: 3+2 facing/unidirectional.
Dimensions: 23.50 x 2.85 m.
Gangways: Within unit only.   **Wheel Arrangement:** 2-B (– B-2) – B-2.
Doors: Twin-leaf swing plug.   **Maximum Speed:** 75 m.p.h.
Multiple Working: Classes 165, 166, 168.

58801–58812/58814–58822. 58873–58878. DMCL. Dia. DP319. Lot No. 31087
BREL York 1990. 16/72 1T. 37.0 t.
58813. DMSL. Dia. DP286. Lot No. 31087 BREL York 1990. –/82 1T. 37.0 t.
58823–58833. DMCL. Dia. DP320. Lot No. 31089 BREL York 1991–92. 24/60 1T.
37.0 t.
MS. Dia. DR208. Lot No. 31090 BREL York 1991–92. –/106. 37.0 t.
DMS. Dia. DP253. Lot No. 31088 BREL York 1991–92. –/98. 37.0 t.

Notes:

165 006–039 are fitted with tripcocks for working over London Underground
tracks between Harrow-on-the-Hill and Amersham.
58813 has been converted to a DMSL with 2+2 seating.
Chiltern Railways units were starting to undergo a refurbishment programme as
this book closed for press and all first class vehicles are to be declassified from
5 January 2003.

165 001	**NT** A *TT*	RG	58801	58834
165 002	**NT** A *TT*	RG	58802	58835
165 003	**NT** A *TT*	RG	58803	58836
165 004	**NT** A *TT*	RG	58804	58837
165 005	**NT** A *TT*	RG	58805	58838
165 006	**NT** A *CR*	AL	58806	58839
165 007	**NT** A *CR*	AL	58807	58840

165 008	NT	A	CR	AL	58808		58841
165 009	NT	A	CR	AL	58809		58842
165 010	NT	A	CR	AL	58810		58843
165 011	NT	A	CR	AL	58811		58844
165 012	NT	A	CR	AL	58812		58845
165 013	CR	A	CR	AL	58813		58846
165 014	NT	A	CR	AL	58814		58847
165 015	NT	A	CR	AL	58815		58848
165 016	NT	A	CR	AL	58816		58849
165 017	NT	A	CR	AL	58817		58850
165 018	NT	A	CR	AL	58818		58851
165 019	NT	A	CR	AL	58819		58852
165 020	NT	A	CR	AL	58820		58853
165 021	NT	A	CR	AL	58821		58854
165 022	NT	A	CR	AL	58822		58855
165 023	NT	A	CR	AL	58873		58867
165 024	NT	A	CR	AL	58874		58868
165 025	NT	A	CR	AL	58875		58869
165 026	NT	A	CR	AL	58876		58870
165 027	NT	A	CR	AL	58877		58871
165 028	NT	A	CR	AL	58878		58872
165 029	NT	A	CR	AL	58823	55404	58856
165 030	NT	A	CR	AL	58824	55405	58857
165 031	NT	A	CR	AL	58825	55406	58858
165 032	NT	A	CR	AL	58826	55407	58859
165 033	NT	A	CR	AL	58827	55408	58860
165 034	NT	A	CR	AL	58828	55409	58861
165 035	NT	A	CR	AL	58829	55410	58862
165 036	NT	A	CR	AL	58830	55411	58863
165 037	NT	A	CR	AL	58831	55412	58864
165 038	NT	A	CR	AL	58832	55413	58865
165 039	NT	A	CR	AL	58833	55414	58866

CLASS 165/1   NETWORK TURBO   BREL

DMCL–DMS or DMCL–MS–DMS.

Construction: Welded aluminium.
Engines: One Perkins 2006-TWH of 260 kW (350 h.p.) at 1900 r.p.m.
Bogies: BREL P3-17 (powered), BREL T3-17 (non-powered).
Couplers: BSI.
Seating Layout: 1: 2+2 facing, 2: 3+2 facing/unidirectional.
Dimensions: 23.50 x 2.85 m.
Gangways: Within unit only.   **Wheel Arrangement:** 2-B (– B-2) – B-2.
Doors: Twin-leaf swing plug.   **Maximum Speed:** 90 m.p.h.
Multiple Working: Classes 165, 166, 168.

58953–58969. DMCL. Dia. DP320. Lot No. 31098 BREL York 1992. 16/66 1T. 37.0 t.
58879–58898. DMCL. Dia. DP319. Lot No. 31096 BREL York 1992. 16/72 1T. 37.0 t.
MS. Dia. DR208. Lot No. 31099 BREL 1992. –/106. 37.0 t.
DMS. Dia. DP253. Lot No. 31097 BREL 1992. –/98. 37.0 t.

165 101	TT	A	TT	RG	58953	55415	58916
165 102	TT	A	TT	RG	58954	55416	58917
165 103	TT	A	TT	RG	58955	55417	58918
165 104	TT	A	TT	RG	58956	55418	58919
165 105	TT	A	TT	RG	58957	55419	58920
165 106	TT	A	TT	RG	58958	55420	58921
165 107	TT	A	TT	RG	58959	55421	58922
165 108	TT	A	TT	RG	58960	55422	58923
165 109	TT	A	TT	RG	58961	55423	58924
165 110	TT	A	TT	RG	58962	55424	58925
165 111	TT	A	TT	RG	58963	55425	58926
165 112	TT	A	TT	RG	58964	55426	58927
165 113	TT	A	TT	RG	58965	55427	58928
165 114	NT	A	TT	RG	58966	55428	58929
165 116	NT	A	TT	RG	58968	55430	58931
165 117	NT	A	TT	RG	58969	55431	58932
165 118	NT	A	TT	RG	58879	58933	
165 119	NT	A	TT	RG	58880	58934	
165 120	NT	A	TT	RG	58881	58935	
165 121	NT	A	TT	RG	58882	58936	
165 122	NT	A	TT	RG	58883	58937	
165 123	NT	A	TT	RG	58884	58938	
165 124	NT	A	TT	RG	58885	58939	
165 125	NT	A	TT	RG	58886	58940	
165 126	NT	A	TT	RG	58887	58941	
165 127	NT	A	TT	RG	58888	58942	
165 128	NT	A	TT	RG	58889	58943	
165 129	NT	A	TT	RG	58890	58944	
165 130	NT	A	TT	RG	58891	58945	
165 131	NT	A	TT	RG	58892	58946	
165 132	NT	A	TT	RG	58893	58947	
165 133	NT	A	TT	RG	58894	58948	
165 134	NT	A	TT	RG	58895	58949	
165 135	NT	A	TT	RG	58896	58950	
165 136	NT	A	TT	RG	58897	58951	
165 137	NT	A	TT	RG	58898	58952	

CLASS 166   NETWORK EXPRESS TURBO   ABB

DMCL (A)–MS–DMCL (B). Built for Paddington–Oxford/Newbury services. Air conditioned.

Construction: Welded aluminium.
Engines: One Perkins 2006-TWH of 260 kW (350 h.p.) at 1900 r.p.m.
Bogies: BREL P3-17 (powered), BREL T3-17 (non-powered).
Couplers: BSI.
Seating Layout: 1: 2+2 facing, 2: 3+2 facing/unidirectional. 20 standard class seats in 2+2 format in DMCL(B).
Dimensions: 23.50 x 2.85 m.

Gangways: Within unit only. **Wheel Arrangement:** 2-B – B-2 – B-2.
Doors: Twin-leaf swing plug. **Maximum Speed:** 90 m.p.h.
Multiple Working: Classes 165, 166, 168.

DMCL (A). Dia. DP321. Lot No. 31116 ABB York 1992–3. 16/75 1T. 40.62 t.
MS. Dia. DR209. Lot No. 31117 ABB York 1992–93. –/96. 38.04 t.
DMCL (B). Dia. DP321. Lot No. 31116 ABB York 1992–93. 16/72 1T. 40.64 t.

166 201	TT	A	TT	RG	58101	58601	58122
166 202	TT	A	TT	RG	58102	58602	58123
166 203	TT	A	TT	RG	58103	58603	58124
166 204	TT	A	TT	RG	58104	58604	58125
166 205	TT	A	TT	RG	58105	58605	58126
166 206	TT	A	TT	RG	58106	58606	58127
166 207	TT	A	TT	RG	58107	58607	58128
166 208	TT	A	TT	RG	58108	58608	58129
166 209	TT	A	TT	RG	58109	58609	58130
166 210	TT	A	TT	RG	58110	58610	58131
166 211	TT	A	TT	RG	58111	58611	58132
166 212	TT	A	TT	RG	58112	58612	58133
166 213	TT	A	TT	RG	58113	58613	58134
166 214	TT	A	TT	RG	58114	58614	58135
166 215	TT	A	TT	RG	58115	58615	58136
166 216	TT	A	TT	RG	58116	58616	58137
166 217	TT	A	TT	RG	58117	58617	58138
166 218	TT	A	TT	RG	58118	58618	58139
166 219	TT	A	TT	RG	58119	58619	58140
166 220	TT	A	TT	RG	58120	58620	58141
166 221	TT	A	TT	RG	58121	58621	58142

CLASS 168 CLUBMAN ADTRANZ/BOMBARDIER

DMSL (A)–MSL–MS–DMSL (B). Air conditioned.

Construction: Welded aluminium bodies with bolt-on steel ends.
Engines: One MTU 6R183TD13H of 315 kW (422 h.p.) at 1900 r.p.m.
Transmission: Hydraulic. Voith T211rzze to ZF final drive.
Bogies: One Adtranz P3–23 and one BREL T3–23 per car.
Couplers: BSI.
Seating Layout: 2+2 facing/unidirectional.
Dimensions: 23.62 x 2.69 m.
Gangways: Within unit only. **Wheel Arrangement:** 2-B (– B-2 – B-2) – B-2.
Doors: Twin-leaf swing plug. **Maximum Speed:** 100 m.p.h.
Multiple Working: Classes 165, 166, 168.
Fitted with tripcocks for working over London Underground tracks between
Harrow-on-the-Hill and Amersham.

Class 168/0. Original Design.

58151–58155. DMSL(A). Dia. DP270. Adtranz Derby 1997–98. –/60 1TD 1W. 43.7 t.
58651–58655. MSL. Dia. DR211. Adtranz Derby 1998. –/73 1T. 41.0 t.
58451–58455. MS. Dia. DR211. Adtranz Derby 1998. –/77. 40.5 t.
58251–58255. DMSL(B). Dia. DP270. Adtranz Derby 1998. –/66 1T. 43.6 t.

Note: 58451–5 formerly numbered 58656–60 in 168 106–110.

168 001	**CR**	P	*CR*	AL	58151	58451	58651	58251
168 002	**CR**	P	*CR*	AL	58152	58452	58652	58252
168 003	**CR**	P	*CR*	AL	58153	58453	58653	58253
168 004	**CR**	P	*CR*	AL	58154	58454	58654	58254
168 005	**CR**	P	*CR*	AL	58155	58455	58655	58255

Class 168/1. These units are effectively Class 170s.

58156–58163. DMSL(A). Dia. DP280. Adtranz Derby 2000. –/59 1TD 2W. 43.7 t.
58164–58167. DMSL(A). Dia. DP280. Bombardier Derby 2003. –/59 1TD 2W. 43.7 t.
58456–58460. MS. Dia. DR211. Bombardier Derby 2002. –/76. 40.5 t.
58461–58463. MS. Dia. DR211. Bombardier Derby 2003. –/76. 42.4 t.
58464. MS. Dia. DR211. Bombardier Derby 2003. –/76. 42.4 t.
58756–58757. MSL. Dia. DR211. Bombardier Derby 2002. –/73 1T. 41.0 t.
58256–58263. DMSL(B). Dia. DP281. Adtranz Derby 2000. –/69 1T. 43.6 t.
58264–58267. DMSL(B). Dia. DP280. Bombardier Derby 2003. –/69 1T. 43.6 t.

Note: 58461–3 have been renumbered from 58661–63.
58756 and 58757 have been renumbered from 58656 and 58657.

168 106	**CR**	P	*CR*	AL	58156	58456	58756	58256
168 107	**CR**	P	*CR*	AL	58157	58457	58757	58257
168 108	**CR**	P	*CR*	AL	58158	58458		58258
168 109	**CR**	P	*CR*	AL	58159	58459		58259
168 110	**CR**	P	*CR*	AL	58160	58460		58260
168 111	**CR**	H	*CR*	AL	58161	58461		58261
168 112	**CR**	H	*CR*	AL	58162	58462		58262
168 113	**CR**	H	*CR*	AL	58163	58463		58263
168 114	**CR**	P			58164	58464		58264
168 115	**CR**	P			58165			58265
168 116	**CR**	P			58166			58266
168 117	**CR**	P			58167			58267

CLASS 170 TURBOSTAR ADTRANZ/BOMBARDIER

Various formations. Air conditioned.

Construction: Welded aluminium bodies with bolt-on steel ends.
Engines: One MTU 6R183TD13H of 315 kW (422 h.p.) at 1900 r.p.m.
Transmission: Hydraulic. Voith T211rzze to ZF final drive.
Bogies: One Adtranz P3–23 and one BREL T3–23 per car.
Couplers: BSI.
Seating Layout: 1: 2+1 facing/unidirectional (2+2 in first class in Class 170/1 end cars). 2: 2+2.
Dimensions: 23.62 x 2.69 m.
Gangways: Within unit only.   **Wheel Arrangement:** 2-B (– B-2) – B-2.
Doors: Twin-leaf swing plug.   **Maximum Speed:** 100 m.p.h.
Multiple Working: Classes 150, 153, 155, 156, 158, 159, 170.

Class 170/1. Midland Mainline Units. DMCL–MCRMB–DMCL or DMCL–DMCL.

DMCL (A). Dia. DP324. Adtranz Derby 1998–1999. 12/45 1TD 2W. 45.19 t.

MCRMB. Dia. DR301. Adtranz Derby 2001. 21/22 and bar. 43.00 t.
DMCL (B). Dia. DP325. Adtranz Derby 1998–1999. 12/52 1T. Catering point. 45.22 t
Note: The DMCL (A) and DMCL (B) in the 3-car units have had their former first class end sections declassified.

170 101	**MM**	P	*MM*	DY	50101	55101	79101
170 102	**MM**	P	*MM*	DY	50102	55102	79102
170 103	**MM**	P	*MM*	DY	50103	55103	79103
170 104	**MM**	P	*MM*	DY	50104	55104	79104
170 105	**MM**	P	*MM*	DY	50105	55105	79105
170 106	**MM**	P	*MM*	DY	50106	55106	79106
170 107	**MM**	P	*MM*	DY	50107	55107	79107
170 108	**MM**	P	*MM*	DY	50108	55108	79108
170 109	**MM**	P	*MM*	DY	50109	55109	79109
170 110	**MM**	P	*MM*	DY	50110	55110	79110
170 111	**MM**	P	*MM*	DY	50111		79111
170 112	**MM**	P	*MM*	DY	50112		79112
170 113	**MM**	P	*MM*	DY	50113		79113
170 114	**MM**	P	*MM*	DY	50114		79114
170 115	**MM**	P	*MM*	DY	50115		79115
170 116	**MM**	P	*MM*	DY	50116		79116
170 117	**MM**	P	*MM*	DY	50117		79117

Class 170/2. Anglia Railways 3-car Units. DMCL–MSLRB–DMSL.

DMCL. Dia. DP326. Adtranz Derby 1999. 30/3 1TD 2W. 44.30 t.
MSLRB. Dia. DR212. Adtranz Derby 1999. –/58 1T. Buffet and guard's office 42.76 t.
DMSL. Dia. DP274. Adtranz Derby 1999. –/66 1T. 44.70 t.

170 201	**AN**	P	*AR*	NC	50201	56201	79201
170 202	**AN**	P	*AR*	NC	50202	56202	79202
170 203	**AN**	P	*AR*	NC	50203	56203	79203
170 204	**AN**	P	*AR*	NC	50204	56204	79204
170 205	**AN**	P	*AR*	NC	50205	56205	79205
170 206	**AN**	P	*AR*	NC	50206	56206	79206
170 207	**AN**	P	*AR*	NC	50207	56207	79207
170 208	**AN**	P	*AR*	NC	50208	56208	79208

Class 170/2. Anglia Railways 2-car Units. DMSL–DMCL.

DMSL. Dia. DP287. Bombardier Derby 2002. –/57 1TD 2W. 44.30 t.
DMCL. Dia. DP274. Bombardier Derby 2002. 9/53 1T. 44.70 t.

170 270	**AN**	P	*AR*	NC	50270	79270
170 271	**AN**	P	*AR*	NC	50271	79271
170 272	**AN**	P	*AR*	NC	50272	79272
170 273	**AN**	P	*AR*	NC	50273	79273

Class 170/3. South West Trains Units. DMCL–DMCL.

DMCL(A). Dia. DP329. Adtranz Derby 2000. 9/43 1TD 2W. 45.80 t.
DMCL(B). Dia. DP330. Adtranz Derby 2000. 9/53 1T. 45.80 t.

170 301	**SW**	P	*SW*	SA	50301	79301
170 302	**SW**	P	*SW*	SA	50302	79302
170 303	**SW**	P	*SW*	SA	50303	79303

170 304	**SW**	P	*SW*	SA	50304	79304
170 305	**SW**	P	*SW*	SA	50305	79305
170 306	**SW**	P	*SW*	SA	50306	79306
170 307	**SW**	P	*SW*	SA	50307	79307
170 308	**SW**	P	*SW*	SA	50308	79308

Class 170/3. Hull Trains 3-car Units. On order. DMCL–MSLRB–DMSL.

DMCL. Dia. DP3??. Bombardier Derby 2003. 30/3 1TD 2W. 44.30 t.
MSLRB. Dia. DR2??. Bombardier Derby 2003. –/58 1T. Buffet and guard's office 42.76 t.
DMSL. Dia. DP2??. Bombardier Derby 2003. –/66 1T. 44.70 t.

170 393	P	50393	56393	79393
170 394	P	50394	56394	79394
170 395	P	50395	56395	79395
170 396	P	50396	56396	79396

Class 170/3. Porterbrook Units. On hire to various operators. DMCL–MC–DMCL or DMCL–DMCL.

DMCL(A). Dia. DP288 or DP329 (170 399). Bombardier Derby 2001–2002. –/56 1TD 1W (9/43 1TD 2W 170 399). 45.40 t.
MC. Dia. DR302. Bombardier Derby 2002. 12/40. 41.60 t.
DMCL(B). Dia. DP289 or DP330 (170 399). Bombardier Derby 2001–2002. –/66 1T (9/53 1T 170 399). 45.80 t.

Advertising Liveries:
170 397 "Q-Jump". Sky blue with purple doors.
170 398 is all over white with large bodyside "BOMBARDIER" lettering.

170 397	**AL**	P	*MM*	DY	50397	56397	79397
170 398	**AL**	P	*CT*	TS	50398	56398	79398
170 399	**P**	P	*CT*	TS	50399		79399

Class 170/4. ScotRail Units. DMCL–MS–DMCL.

DMCL(A). Dia. DP329. Adtranz Derby 1999–2001. 9/43 1TD 2W. 45.80 t.
MS. Dia. DR213. Adtranz Derby 1999–2001. –/76. 43.00 t.
DMCL(B). Dia. DP330. Adtranz Derby 1999–2001. 9/53 1T. 45.80 t.

170 401	**SR**	P	*SR*	HA	50401	56401	79401
170 402	**SR**	P	*SR*	HA	50402	56402	79402
170 403	**SR**	P	*SR*	HA	50403	56403	79403
170 404	**SR**	P	*SR*	HA	50404	56404	79404
170 405	**SR**	P	*SR*	HA	50405	56405	79405
170 406	**SR**	P	*SR*	HA	50406	56406	79406
170 407	**SR**	P	*SR*	HA	50407	56407	79407
170 408	**SR**	P	*SR*	HA	50408	56408	79408
170 409	**SR**	P	*SR*	HA	50409	56409	79409
170 410	**SR**	P	*SR*	HA	50410	56410	79410
170 411	**SR**	P	*SR*	HA	50411	56411	79411
170 412	**SR**	P	*SR*	HA	50412	56412	79412
170 413	**SR**	P	*SR*	HA	50413	56413	79413
170 414	**SR**	P	*SR*	HA	50414	56414	79414
170 415	**SR**	P	*SR*	HA	50415	56415	79415

170 416	**SR**	H	*SR*	HA	50416	56416	79416
170 417	**SR**	H	*SR*	HA	50417	56417	79417
170 418	**SR**	H	*SR*	HA	50418	56418	79418
170 419	**SR**	H	*SR*	HA	50419	56419	79419
170 420	**SR**	H	*SR*	HA	50420	56420	79420
170 421	**SR**	H	*SR*	HA	50421	56421	79421
170 422	**SR**	H	*SR*	HA	50422	56422	79422
170 423	**SR**	H	*SR*	HA	50423	56423	79423
170 424	**SR**	H	*SR*	HA	50424	56424	79424

Class 170/4. ScotRail Units. Additional units (standard class only) for Strathclyde PTE. DMSL–MS–DMSL.

DMSL(A). Dia. DP284. Adtranz Derby 2001. –/55 1TD 2W. 45.80 t.
MS. Dia. DR213. Adtranz Derby 2001. –/76. 43.00 t.
DMSL(B). Dia. DP285. Adtranz Derby 2001. –/67 1T. 45.80 t.

170 470	**SP**	P	*SR*	HA	50470	56470	79470
170 471	**SP**	P	*SR*	HA	50471	56471	79471

Class 170/5. Central Trains 2-car Units. DMSL–DMSL.

DMSL(A). Dia. DP275. Adtranz Derby 1999–2000. –/55 1TD 2W. 45.80 t.
DMSL(B). Dia. DP276. Adtranz Derby 1999–2000. –/67 1T. 46.80 t.

170 501	r	**CT**	P	*CT*	TS	50501	79501
170 502	r	**CT**	P	*CT*	TS	50502	79502
170 503	r	**CT**	P	*CT*	TS	50503	79503
170 504	r	**CT**	P	*CT*	TS	50504	79504
170 505	r	**CT**	P	*CT*	TS	50505	79505
170 506	r	**CT**	P	*CT*	TS	50506	79506
170 507	r	**CT**	P	*CT*	TS	50507	79507
170 508	r	**CT**	P	*CT*	TS	50508	79508
170 509	r	**CT**	P	*CT*	TS	50509	79509
170 510	r	**CT**	P	*CT*	TS	50510	79510
170 511	r	**CT**	P	*CT*	TS	50511	79511
170 512	r	**CT**	P	*CT*	TS	50512	79512
170 513	r	**CT**	P	*CT*	TS	50513	79513
170 514	r	**CT**	P	*CT*	TS	50514	79514
170 515	r	**CT**	P	*CT*	TS	50515	79515
170 516	r	**CT**	P	*CT*	TS	50516	79516
170 517	r	**CT**	P	*CT*	TS	50517	79517
170 518	r	**CT**	P	*CT*	TS	50518	79518
170 519	r	**CT**	P	*CT*	TS	50519	79519
170 520	r	**CT**	P	*CT*	TS	50520	79520
170 521	r	**CT**	P	*CT*	TS	50521	79521
170 522	r	**CT**	P	*CT*	TS	50522	79522
170 523	r	**CT**	P	*CT*	TS	50523	79523

Class 170/6. Central Trains 3-car Units. DMSL–MS–DMSL.

DMSL(A). Dia. DP275. Adtranz Derby 2000. –/55 1TD 2W. 45.80 t.
MS. Dia. DR214. Adtranz Derby 2000. –/74. 43.00 t.
DMSL(B). Dia. DP276. Adtranz Derby 2000. –/67 1T. 46.80 t.

170 630	r	**CT**	P	*CT*	TS	50630	56630	79630
170 631	r	**CT**	P	*CT*	TS	50631	56631	79631
170 632	r	**CT**	P	*CT*	TS	50632	56632	79632
170 633	r	**CT**	P	*CT*	TS	50633	56633	79633
170 634	r	**CT**	P	*CT*	TS	50634	56634	79634
170 635	r	**CT**	P	*CT*	TS	50635	56635	79635
170 636	r	**CT**	P	*CT*	TS	50636	56636	79636
170 637	r	**CT**	P	*CT*	TS	50637	56637	79637
170 638	r	**CT**	P	*CT*	TS	50638	56638	79638
170 639	r	**CT**	P	*CT*	TS	50639	56639	79639

Class 170/7. GoVia South Central Units. On order. DMCL–DMCL.

DMSL. Dia. DPxxx. Bombardier Derby 2003. x/xx 1TD 2W. xx.xx t.
DMCL. Dia. DPxxx. Bombardier Derby 2003. x/xx 1T. xx.xx t.

170 721	**SN**	P	50721	79721
170 722	**SN**	P	50722	79722
170 723	**SN**	P	50723	79723
170 724	**SN**	P	50724	79724
170 725	**SN**	P	50725	79725
170 726	**SN**	P	50726	79726
170 727	**SN**	P	50727	79727
170 728	**SN**	P	50728	79728
170 729	**SN**	P	50729	79729
170 730	**SN**	P	50730	79730
170 731	**SN**	P	50731	79731
170 732	**SN**	P	50732	79732
170 733	**SN**	P	50733	79733
170 734	**SN**	P	50734	79734
170 735	**SN**	P	50735	79735
170 736	**SN**	P	50736	79736
170 737	**SN**	P	50737	79737
170 738	**SN**	P	50738	79738

CLASS 175   CORADIA 1000   ALSTOM

Air Conditioned.

Construction: Steel.
Engines: One Cummins N14 of 335 kW (450 h.p.).
Transmission: Hydraulic. Voith T211rzze to ZF final drive.
Bogies:
Couplers: Scharfenberg.
Seating Layout: 2+2 facing/unidirectional.
Dimensions: 23.71/23.03 x 2.73 m.
Gangways: Within unit only.   **Wheel Arrangement:** 2-B (– B-2) – B-2.
Doors: Single-leaf swing plug.   **Maximum Speed:** 100 m.p.h.
Multiple Working: Classes 175, 180.

Class 175/0. DMSL–DMSL. 2-car units.

DMSL(A). Dia. DP278. Alstom Birmingham 1999–2000. –/54 1TD 2W. 51.00 t.
DMSL(B). Dia. DP279. Alstom Birmingham 1999–2000. –/64 1T. 51.00 t.

175 001	FS	A	NW	CH	50701	79701
175 002	FS	A	NW	CH	50702	79702
175 003	FS	A	NW	CH	50703	79703
175 004	FS	A	NW	CH	50704	79704
175 005	FS	A	NW	CH	50705	79705
175 006	FS	A	NW	CH	50706	79706
175 007	FS	A	NW	CH	50707	79707
175 008	FS	A	NW	CH	50708	79708
175 009	FS	A	NW	CH	50709	79709
175 010	FS	A	NW	CH	50710	79710
175 011	FS	A	NW	CH	50711	79711

Names:

175 004   MENCAP National Colleges Pengwern College
175 008   Valhalla Blackpool Pleasure Beach

Class 175/1. DMSL–MSL–DMSL. 3-car units.

DMSL(A). Dia. DP278. Alstom Birmingham 1999–2001. –/54 1TD 2W. 51.00 t.
MSL. Dia. DR216. Alstom Birmingham 1999–2001. –/68 1T. 43 t 68 1T. 47.50 t.
DMSL(B). Dia. DP279. Alstom Birmingham 1999–2001. –/64 1T. 51.00 t.

175 101	FS	A	NW	CH	50751	56751	79751
175 102	FS	A	NW	CH	50752	56752	79752
175 103	FS	A	NW	CH	50753	56753	79753
175 104	FS	A	NW	CH	50754	56754	79754
175 105	FS	A	NW	CH	50755	56755	79755
175 106	FS	A	NW	CH	50756	56756	79756
175 107	FS	A	NW	CH	50757	56757	79757
175 108	FS	A	NW	CH	50758	56758	79758
175 109	FS	A	NW	CH	50759	56759	79759
175 110	FS	A	NW	CH	50760	56760	79760
175 111	FS	A	NW	CH	50761	56761	79761
175 112	FS	A	NW	CH	50762	56762	79762
175 113	FS	A	NW	CH	50763	56763	79763
175 114	FS	A	NW	CH	50764	56764	79764
175 115	FS	A	NW	CH	50765	56765	79765
175 116	FS	A	NW	CH	50766	56766	79766

Names:

175 103   Mum
175 111   Brief Encounter
175 112   South Lakes Wild Animal Park SUMATRAN TIGER
175 114   Commonwealth Cruiser

CLASS 180   CORADIA 1000   ALSTOM

New units for First Great Western.

Construction: Steel.
Engines: One Cummins QSK19 of 560 kW (750 h.p.) at 2100 r.p.m.
Transmission: Hydraulic. Voith T312br to Voith final drive.
Bogies: Alstom MB2.
Couplers: Scharfenberg.
Seating Layout: 1: 2+1, 2: 2+2 facing/unidirectional.
Dimensions: 23.71/23.03 x 2.73 m.
Gangways: Within unit only.   **Wheel Arrangement:** 2-B – B-2 – B-2 – B-2.
Doors: Single-leaf swing plug.   **Maximum Speed:** 125 m.p.h.
Multiple Working: Classes 175, 180.

DMSL(A). Dia. DP282. Alstom Birmingham 2000–01. –/46 2W 1TD. 53.00 t.
MFL. Dia. DR101. Alstom Birmingham 2000–01. 42/– 1T 1W + catering point. 51.50 t.
MSL. Dia. DR217. Alstom Birmingham 2000–01. –/68 1T. 51.50 t.
MSLRB. Dia. DR218. Alstom Birmingham 2000–01. –/56 1T. 51.50 t.
DMSL(B). Dia. DP283. Alstom Birmingham 2000–01. –/56 1T. 53.00 t.

180 101	**FG**	W			50901	54901	55901	56901	59901
180 102	**FG**	W	GW	OM	50902	54902	55902	56902	59902
180 103	**FG**	W	GW	OM	50903	54903	55903	56903	59903
180 104	**FG**	W	GW	OM	50904	54904	55904	56904	59904
180 105	**FG**	W	GW	OM	50905	54905	55905	56905	59905
180 106	**FG**	W	GW	OM	50906	54906	55906	56906	59906
180 107	**FG**	W	GW	OM	50907	54907	55907	56907	59907
180 108	**FG**	W	GW	OM	50908	54908	55908	56908	59908
180 109	**FG**	W	GW	OM	50909	54909	55909	56909	59909
180 110	**FG**	W	GW	OM	50910	54910	55910	56910	59910
180 111	**FG**	W	GW	OM	50911	54911	55911	56911	59911
180 112	**FG**	W	GW	OM	50912	54912	55912	56912	59912
180 113	**FG**	W	GW	OM	50913	54913	55913	56913	59913
180 114	**FG**	W			50914	54914	55914	56914	59914

2. DIESEL ELECTRIC UNITS

The following features are standard to ex-BR Southern Region diesel-electric multiple unit power cars (Classes 201–207):

Construction: Steel.
Engine: One English Electric 4SRKT Mk. 2 of 450 kW (600 h.p.) at 850 r.p.m.
Main Generator: English Electric EE824.
Traction Motors: Two English Electric EE507 mounted on the inner bogie.

Bogies: SR Mk. 4. (Former EMU TSL vehicles have Commonwealth bogies).
Couplers: Drophead buckeye.
Doors: Manually operated slam.
Brakes: Electro-pneumatic and automatic air.
Maximum Speed: 75 m.p.h.
Multiple Working: Other ex BR Southern Region DEMU vehicles.

CLASS 201/202 PRESERVED 'HASTINGS' UNIT BR

DMBS–2TSL–TSRB–TSL–DMBS.

Preserved unit made up from 2 Class 201 short-frame cars and 2 Class 202 long-frame cars. The 'Hastings' units were made with narrow body-profiles for use on the section between Tonbridge and Battle which had tunnels of restricted loading gauge. These tunnels were converted to single track operation in the 1980s thus allowing standard loading gauge stock to be used. The set also contains a Class 411 EMU trailer (not Hastings line gauge).

Gangways: Within unit only.
Seating Layout: 2+2 facing.
Dimensions: 18.36 x 2.50 m (60000/60501), 20.34 x 2.50 m. (60118/60529) 20.34 x 2.82 m (69337/70262).

60000. DMBS. Dia DB201. Lot No. 30329 Eastleigh 1957. –/22. 54 t.
60501. TSL. Dia DH201. Lot No. 30331 Eastleigh 1957. –/52 2T. 29 t.
70262. TSL (ex Class 411/5 EMU). Dia. DH208. Lot No. 30455 Eastleigh 1958–99. –/64 2T. 33.78 t.
69337. TSRB (ex Class 422 EMU). Dia. DH209. Lot No. 30805 York 1970. –/40. 35 t.
60529. TSL. Dia DH202. Lot No. 30397 Eastleigh 1957. –/60 2T. 30 t.
60118. DMBS. Dia DB202. Lot No. 30395 Eastleigh 1957. –/30. 55 t.

201 001   **G**   HD *SS*   SE   60000 60529 70262 69337 60501 60118

Names:

60000   Hastings | 60118   Tunbridge Wells

CLASS 205/0 (3H)   BR 'HAMPSHIRE'

DMBS–TSL–DTCsoL or DMBS–DTCsoL.

Gangways: Non-gangwayed.
Seating Layout: 3+2 facing or compartments.
Dimensions: 20.33 × 2.82 m (DMBS), 20.28 × 2.82 m (TS), 20.36 × 2.82 m (DTCsoL).

60111/117/154. DMBS. Dia DB203. Lot No. 30332 Eastleigh 1957. –/52. 56 t.
60122–124. DMBS. Dia DB203. Lot No. 30540 Eastleigh 1958–59. –/52. 56 t.
60146–151. DMBS. Dia DB204. Lot No. 30671 Eastleigh 1960–62. –/42. 56 t.
60650–670. TS. Dia DH203. Lot No. 30542 Eastleigh 1958–59. –/104. 30 t.
60673–678. TS. Dia DH203. Lot No. 30672 Eastleigh 1960–62. –/104. 30 t.
60800–60808. DTCsoL. Dia DE301. Lot No. 30333 Eastleigh 1956–57. 13/50 2T. 32 t.
60811. DTCsoL. Dia DE302. Lot No. 30333 Eastleigh 1956–57. 19/50 2T. 32 t.
60820. DTCsoL. Dia DE301. Lot No. 30399 Eastleigh 1957–58. 13/50 2T. 32 t.
60823/824. DTCsoL. Dia DE301. Lot No. 30541 Eastleigh 1958–59. 13/50 2T. 32 t.
60827–832. DTCsoL. Dia DE303. Lot No. 30673 Eastleigh 1960–62. 13/62 2T. (13/60 2T 60827 (DE304), 13/57 2T 60831 (DE305)) 32 t.

205 001	CX	P	SC	SU	60154		60800
205 009	CX	P	SC	SU	60108	60658	60808
205 012	CX	P	SC	SU	60111		60811
205 018	CX	P	SC	SU	60117	60674	60828
205 025	CX	P	SC	SU	60124		60824
205 028	CX	P	SC	SU	60146	60673	60827
205 032	CX	P	SC	SU	60150	60677	60831
205 033	CX	P	SC	SU	60151	60678	60832
Spare	CX	P		ZG	60650		
Spare	N	P		SU	60123		60823
Spare	N	P		SU		60670	
Spare	G	HD		SE	60122	60668	

CLASS 205/2 (3H)   BR 'HAMPSHIRE'

DMBS–TSL (ex Class 411/5 EMU)–DTSL. Refurbished 1980. Fluorescent lighting.

Details as for Class 205/0 except:

Gangways: Within unit only.
Seating Layout: 3+2 facing.

DMBS. Dia. DB203. Lot No. 30332 Eastleigh 1957. –/39. 57 t.
TSL. Dia. DH207. Converted from loco-hauled TS 4059 Lot No. 30149 Swindon 1955–57. –/64 2T. 33.78 t.
DTSL. Dia. DE204. Lot No. 30333 Eastleigh 1957. –/76 2T. 32 t.
Note: This unit operates as a two-car set in winter.

205 205	CX	P	SC	SU	60110	71634	60810

CLASS 207/0 (2D)   BR 'OXTED'

DMBS–DTS (formerly DMBS–TCsoL–DTS).

This class was built for the Oxted line and therefore referred to as 'Oxted' units. They were made with a narrower body-profile which also allowed them to be used through the restricted loading-gauge Somerhill Tunnel between Tonbridge and Grove Junction (Tunbridge Wells). This tunnel was converted to single track operation in the 1980s thus allowing standard loading gauge stock to be used.

Gangways: Non-gangwayed.
Seating Layout: 3+2 facing or compartments.
Dimensions: 20.33 x 2.74 m (DMBS/TCsoL), 20.32 x 2.74 m. (DTS).

DMBS. Dia DB205. Lot No. 30625 Eastleigh 1962. –/42. 56 t.
60616. TCsoL. Dia DH301. Lot No. 30626 Eastleigh 1962. 24/42 1T. 31 t.
60916. DTS. Dia DE201. Lot No. 30627 Eastleigh 1962. –/76. 32 t.

207 017	**CX**	P	*SC*	SU	60142		60916
Spare	**G**	HD		SE	60138	60616	

CLASS 207/2 (3D)   BR 'OXTED'

DMBS–TSL–DTS.
Gangwayed sets with an ex-Class 411 EMU trailer in the centre.

Gangways: Within unit only.
Seating Layout: 2+2 facing.
Dimensions: 20.34 x 2.74 m (DMBS), 20.32 x 2.74 m. (DTS).

DMBS. Dia DB205. Lot No. 30625 Eastleigh 1962. –/40. 56 t.
70286. TSL. Dia. DH206. Lot No. 30455 Eastleigh 1958–59. –/64 2T. 33.78 t.
70547/9. TSL. Dia. DH206. Lot No. 30620 Eastleigh 1960–61 –/64 2T. 33.78 t.
DTS. Dia DE201. Lot No. 30627 Eastleigh 1962. –/75. 32 t.

Note: These units operate as two-car sets in winter.

207 201	**CX**	P	*SC*	SU (S)	60129	70286	60903
207 202	**CX**	P	*SC*	SU	60130	70549	60904
207 203	**CX**	P	*SC*	SU	60127	70547	60901

Names:

207 201 Ashford Fayre | 207 202 Brighton Royal Pavilion

CLASS 220   VOYAGER   BOMBARDIER

DMS–MSRMB–MS–DMF. Units for Virgin Cross-Country.

Construction: Steel.
Engine: Cummins of 750 h.p. (560 kW) at 1800 r.p.m.
Transmission: Two Alstom Onix 800 three-phase traction motors of 275 kW.
Braking: Rheostatic and electro-pneumatic.
Bogies: Bombardier B5005.
Couplers: Dellner.
Seating Layout: 1: 2+1 facing/unidirectional, 2: 2+2 mainly unidirectional.
Dimensions: 23.85/22.82 x 2.73 m.
Gangways: Within unit only.
Wheel Arrangement: 1A-A1 – 1A-A1 – 1A-A1 – 1A-A1.
Doors: Single-leaf swing plug.
Maximum Speed: 125 m.p.h.
Multiple Working: Classes 220, 221, 222

DMS. Dia DC201. Bombardier Brugge/Wakefield 2000–01. –/42 1TD 1W. 48.10 t.
MSRMB. Dia. DD201. Bombardier Brugge/Wakefield 2000–01. –/58. 48.00 t.
MS. Dia. DD202. Bombardier Brugge/Wakefield 2000–01. –/62 1TD 1W. 45.00 t.
DMF. Dia DC101. Bombardier Brugge/Wakefield 2000–01. 26/– 1TD 1W. 44.50 t.

220 001	VT	HX	VX	CZ	60301	60701	60201	60401
220 002	VT	HX	VX	CZ	60302	60702	60202	60402
220 003	VT	HX	VX	CZ	60303	60703	60203	60403
220 004	VT	HX	VX	CZ	60304	60704	60204	60404
220 005	VT	HX	VX	CZ	60305	60705	60205	60405
220 006	VT	HX	VX	CZ	60306	60706	60206	60406
220 007	VT	HX	VX	CZ	60307	60707	60207	60407
220 008	VT	HX	VX	CZ	60308	60708	60208	60408
220 009	VT	HX	VX	CZ	60309	60709	60209	60409
220 010	VT	HX	VX	CZ	60310	60710	60210	60410
220 011	VT	HX	VX	CZ	60311	60711	60211	60411
220 012	VT	HX	VX	CZ	60312	60712	60212	60412
220 013	VT	HX	VX	CZ	60313	60713	60213	60413
220 014	VT	HX	VX	CZ	60314	60714	60214	60414
220 015	VT	HX	VX	CZ	60315	60715	60215	60415
220 016	VT	HX	VX	CZ	60316	60716	60216	60416
220 017	VT	HX	VX	CZ	60317	60717	60217	60417
220 018	VT	HX	VX	CZ	60318	60718	60218	60418
220 019	VT	HX	VX	CZ	60319	60719	60219	60419
220 020	VT	HX	VX	CZ	60320	60720	60220	60420
220 021	VT	HX	VX	CZ	60321	60721	60221	60421
220 022	VT	HX	VX	CZ	60322	60722	60222	60422
220 023	VT	HX	VX	CZ	60223	60723	60223	60423
220 024	VT	HX	VX	CZ	60224	60724	60224	60424
220 025	VT	HX	VX	CZ	60225	60725	60225	60425
220 026	VT	HX	VX	CZ	60226	60726	60226	60426
220 027	VT	HX	VX	CZ	60227	60727	60227	60427
220 028	VT	HX	VX	CZ	60228	60728	60228	60428

220 029	**VT**	HX	*VX*	CZ	60229	60729	60229	60429
220 030	**VT**	HX	*VX*	CZ	60330	60730	60230	60430
220 031	**VT**	HX	*VX*	CZ	60331	60731	60231	60431
220 032	**VT**	HX	*VX*	CZ	60332	60732	60232	60432
220 033	**VT**	HX	*VX*	CZ	60333	60733	60233	60433
220 034	**VT**	HX	*VX*	CZ	60334	60734	60234	60434

Names (carried on MSRMB):

220 001	Maiden Voyager	220 019	Mersey Voyager
220 002	Forth Voyager	220 020	Wessex Voyager
220 003	Solent Voyager	220 021	Blackpool Voyager
220 004	Cumbrian Voyager	220 022	Brighton Voyager
220 005	Guildford Voyager	220 023	Mancunian Voyager
220 006	Clyde Voyager	220 024	Sheffield Voyager
220 007	Thames Voyager	220 025	Severn Voyager
220 008	Draig Gymreig/Welsh Dragon	220 026	Stagecoach Voyager
220 009	Gatwick Voyager	220 027	Avon Voyager
220 010	Ribble Voyager	220 028	Blackcountry Voyager
220 011	Tyne Voyager	220 029	Cornish Voyager/
220 012	Lanarkshire Voyager		Vyajer Kernewek
220 013	Gwibiwr De Cymru/	220 030	Devon Voyager
	South Wales Voyager	220 031	Tay Voyager
220 014	South Yorkshire Voyager	220 032	Grampian Voyager
220 015	Solway Voyager	220 033	Fife Voyager
220 016	Midland Voyager	220 034	Yorkshire Voyager
220 017	BOMBARDIER Voyager		
220 018	Central News		

CLASS 221   SUPER VOYAGER   BOMBARDIER

DMS–MSRMB(–MS)–MS–DMF. Tilting units for Virgin Cross-Country.

Construction: Steel.
Engine: Cummins of 750 h.p. (560 kW) at 1800 r.p.m.
Transmission: Two Alstom Onix 800 three-phase traction motors of 275 kW.
Braking: Rheostatic and electro-pneumatic.
Bogies: Bombardier HVP.
Couplers: Dellner.
Seating Layout: 1: 2+1 facing/unidirectional, 2: 2+2 mainly unidirectional.
Dimensions: 23.85/22.82 x 2.73 m.
Gangways: Within unit only.
Wheel Arrangement: 1A-A1 – 1A-A1 – 1A-A1 (– 1A-A1) – 1A-A1.
Doors: Single-leaf swing plug.
Maximum Speed: 125 m.p.h.
Multiple Working: Classes 220, 221, 222.

DMS. Dia DF201. Bombardier Brugge/Wakefield 2001–02. –/42 1TD 1W. 58.30 t.
MSRMB. Dia. DG201. Bombardier Brugge/Wakefield 2001–02. –/58. 58.00 t.
MS. Dia. DG202. Bombardier Brugge/Wakefield 2001–02. –/62 1TD 1W. 55.80 t.
DMF. Dia DF101. Bombardier Brugge/Wakefield 2001–02. 26/– 1TD 1W. 54.90 t.

221 101	VT	HX			60351	60751	60951	60851	60451
221 102	VT	HX	VX	CZ	60352	60782	60952	60852	60452
221 103	VT	HX	VX	CZ	60353	60783	60953	60853	60453
221 104	VT	HX	VX	CZ	60354	60785	60954	60854	60454
221 105	VT	HX	VX	CZ	60355	60755	60955	60855	60455
221 106	VT	HX	VX	CZ	60356	60756	60956	60856	60456
221 107	VT	HX	VX	CZ	60357	60757	60957	60857	60457
221 108	VT	HX	VX	CZ	60358	60758	60958	60858	60458
221 109	VT	HX	VX	CZ	60359	60759	60959	60859	60459
221 110	VT	HX	VX	CZ	60360	60760	60960	60860	60460
221 111	VT	HX	VX	CZ	60361	60761	60961	60861	60461
221 112	VT	HX	VX	CZ	60362	60762	60962	60862	60462
221 113	VT	HX	VX	CZ	60363	60763	60963	60863	60463
221 114	VT	HX	VX	CZ	60364	60764	60964	60864	60464
221 115	VT	HX	VX	CZ	60365	60765	60965	60865	60465
221 116	VT	HX	VX	CZ	60366	60766	60966	60866	60466
221 117	VT	HX	VX	CZ	60367	60767	60967	60867	60467
221 118	VT	HX	VX	CZ	60368	60768	60968	60868	60468
221 119	VT	HX	VX	CZ	60369	60769	60969	60869	60469
221 120	VT	HX	VX	CZ	60370	60770	60970	60870	60470
221 121	VT	HX	VX	CZ	60371	60771	60971	60871	60471
221 122	VT	HX	VX	CZ	60372	60772	60972	60872	60472
221 123	VT	HX	VX	CZ	60373	60773	60973	60873	60473
221 124	VT	HX	VX	CZ	60374	60774	60974	60874	60474
221 125	VT	HX	VX	CZ	60375	60775	60975	60875	60475
221 126	VT	HX	VX	CZ	60376	60776	60976	60876	60476
221 127	VT	HX	VX	CZ	60377	60777	60977	60877	60477
221 128	VT	HX	VX	CZ	60378	60778	60978	60878	60478
221 129	VT	HX	VX	CZ	60379	60779	60979	60879	60479
221 130	VT	HX	VX	CZ	60380	60780	60980	60880	60480
221 131	VT	HX	VX	CZ	60381	60781	60981	60881	60481
221 132	VT	HX	VX	CZ	60382	60782	60982	60882	60482
221 133	VT	HX	VX	CZ	60383	60783	60983	60883	60483
221 134	VT	HX	VX	CZ	60384	60784	60984	60884	60484
221 135	VT	HX	VX	CZ	60385	60785	60985	60885	60485
221 136	VT	HX	VX	CZ	60386	60786	60986	60886	60486
221 137	VT	HX	VX	CZ	60387	60787	60987	60887	60487
221 138	VT	HX	VX	CZ	60388	60788	60988	60888	60488
221 139	VT	HX	VX	CZ	60389	60789	60989	60889	60489
221 140	VT	HX	VX	CZ	60390	60790	60990	60890	60490
221 141	VT	HX	VX	CZ	60391	60791	60991		60491
221 142	VT	HX	VX	CZ	60392	60792	60992		60492
221 143	VT	HX	VX	CZ	60393	60793	60993		60493
221 144	VT	HX	VX	CZ	60394	60794	60994		60494

Names (carried on MSRMB):

221 101	Louis Bleriot		221 123	Henry Hudson
221 102	John Cabot		221 124	Charles Lindbergh
221 103	Christopher Columbus		221 125	Henry the Navigator
221 104	Sir John Franklin		221 126	Captain Robert Scott
221 105	William Baffin		221 127	Wright Brothers
221 106	Willem Barents		221 128	Captain John Smith
221 107	Sir Martin Frobisher		221 129	George Vancouver
221 108	Sir Ernest Shackleton		221 130	Michael Palin
221 109	Marco Polo		221 131	Edgar Evans
221 110	James Cook		221 132	William Speirs Bruce
221 111	Roald Amundsen		221 133	Alexander Selkirk
221 112	Ferdinand Magellan		221 134	Mary Kingsley
221 113	Sir Walter Raleigh		221 135	Donald Campbell
221 114	Sir Francis Drake		221 136	Yuri Gagarin
221 115	Sir Francis Chichester		221 137	Mayflower Pilgrims
221 116	David Livingstone		221 138	Thor Heyerdahl
221 117	Sir Henry Morton-Stanley		221 139	Leif Erikson
221 118	Mungo Park		221 140	Vasco da Gama
221 119	Amelia Earhart		221 141	Amerigo Vespucci
221 120	Amy Johnson		221 142	Mathew Flinders
221 121	Charles Darwin		221 143	Auguste Picard
221 122	Doctor Who		221 144	Prince Madoc

CLASS 222   MERIDIAN   BOMBARDIER

Various formations. New 9-car and 4-car units on order for Midland Mainline.

Construction: Steel.
Engine: Cummins of 750 h.p. (560 kW) at 1800 r.p.m.
Transmission: Two Alstom Onix 800 three-phase traction motors of 275 kW.
Braking: Rheostatic and electro-pneumatic.
Bogies: Bombardier B5005.
Couplers: Dellner.
Seating Layout: 1: 2+1, 2: 2+2 facing/unidirectional.
Dimensions: 23.85/22.82 x 2.73 m.
Gangways: Within unit only.
Wheel Arrangement: All cars 1A-A1.
Doors: Single-leaf swing plug.
Maximum Speed: 125 m.p.h.
Multiple Working: Classes 220, 221, 222.

DMRFO. Dia DD1 . Bombardier Brugge/Wakefield 2002–04. 22/– 1TD 1W. . t.
MFO. Dia. DC1 . Bombardier Brugge/Wakefield 2002–04. 42/– 1T. . t.
MCO. Dia. DC3 . Bombardier Brugge/Wakefield 2002–04. 28/20 1T. . t.
MSORMB. Dia. DC2 . Bombardier Brugge/Wakefield 2002–04. –/62. . t.
MSO. Dia. DC2 . Bombardier Brugge/Wakefield 2002–04. –/70 1T. . t.
DMSO. Dia DD2 . Bombardier Brugge/Wakefield 2002–04. –/36 1TD 2W. . t.

222 001–007. DMRFO–MFO–MFO–MSO–MSO–MSORMB–MSO–MSO–DMSO.
9-car units.

222 001	**MN**	H	60161	60531	60541	60551	60561
			60621	60341	60441	60241	
222 002	**MN**	H	60162	60532	60542	60552	60562
			60622	60342	60442	60242	
222 003	**MN**	H	60163	60533	60543	60553	60563
			60623	60343	60443	60243	
222 004	**MN**	H	60164	60534	60544	60554	60564
			60624	60344	60444	60244	
222 005	**MN**	H	60165	60535	60545	60555	60565
			60625	60345	60445	60245	
222 006	**MN**	H	60166	60536	60546	60556	60566
			60626	60346	60446	60246	
222 007	**MN**	H	60167	60537	60547	60557	60567
			60627	60347	60447	60247	

222 008–023. DMRFO–MCO–MSORMB–DMSO. 4-car units.

222 008	**MN**	H	60168	60918	60628	60248
222 009	**MN**	H	60169	60919	60629	60249
222 010	**MN**	H	60170	60920	60630	60250
222 011	**MN**	H	60171	60921	60631	60251
222 012	**MN**	H	60172	60922	60632	60252
222 013	**MN**	H	60173	60923	60633	60253
222 014	**MN**	H	60174	60922	60634	60254
222 015	**MN**	H	60175	60925	60635	60255
222 016	**MN**	H	60176	60926	60636	60256
222 017	**MN**	H	60177	60927	60637	60257
222 018	**MN**	H	60178	60928	60638	60258
222 019	**MN**	H	60179	60929	60639	60259
222 020	**MN**	H	60180	60930	60640	60260
222 021	**MN**	H	60181	60931	60641	60261
222 022	**MN**	H	60182	60932	60642	60262
222 023	**MN**	H	60183	60933	60643	60263

▲ Newly repainted Thames Trains liveried 165 106 is seen near Salfords with the 07.12 Reading–Gatwick Airport on 23/06/02. **Alex Dasi-Sutton**

▼ 166 213 passes Shallesbrook on the Great Western main line with a Reading–Paddington service on 01/09/01. Note the variation in the Thames Trains Class 165 and 166 liveries. **Rodney Lissenden**

▲ Chiltern Railways liveried 168 108 passes Tyseley on 28/07/02 with a Marylebone–Birmingham Moor Street service. **Jason Rogers**

▼ Central Trains' 170s are one of the most travelled classes of DMU. Most services on the Liverpool–Norwich corridor are booked for the class and on 12/09/02 170 507 is seen pausing at Wymondham with the 08.52 Liverpool–Norwich. **Robert Pritchard**

Midland Mainline liveried 170 110 leaves Wellingborough on 24/09/02 with the 08.52 Nottingham–St. Pancras.

Dick Crane

▲ 175 102 is seen leaving Abergele & Pensarn with a Llandudno–Chester service on 31/08/02. **Jason Rogers**

▼ "Adelante" unit 180 107 passes Manor Farm on the GWML with the 15.30 Paddington–Cardiff on 14/08/02. **Anthony Kay**

"Hastings" DEMU Class 201 No. 201 001 pulls away from Crowhurst whilst working a Tonbridge–Hastings service during the line's 150th anniversary day - 01/09/01.

Ian Feather

▲ South Central operated 205 025, still in Connex livery, arrives at Rye on an Ashford–Hastings service on 02/08/02. These units are to be replaced on this line and on the Uckfield line by Class 170s. **John Chalcraft**

▼ 3-car 207 203 is seen here approaching May Cross with the 12.52 Ashford–Hastings service. **Ian Feather**

▲ All of Virgin Cross-Country's Class 220 and 221 "Voyager" fleet is now in operation, working services the length and breadth of the country. On 28/07/02 220 029 "Cornish Voyager" and 220 004 "New Dawn" (since re-named "Cumbrian Voyager") pass Whitacre Junction with the diverted 10.04 Birmingham–Manchester.
Hugh Ballantyne

▼ 221 140 "Vasco da Gama" leaves Sheffield on 26/10/02 with the 09.58 Newcastle–Cardiff Central.
Robert Pritchard

▲ Class 960 (ex-class 121) "Bubble-car" 977 968 is seen at Rugby on 01/05/02. This unit is operated by Eurailscout GB and is used as a track recording unit.
Andy Flowers

▼ BR maroon liveried sandite "Bubble-car" No. 960 010 (55024) is seen at Marylebone on 09/10/01. **Paul Chancellor**

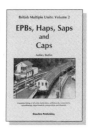

PLATFORM 5 MAIL ORDER

RAILWAY TRACK DIAGRAMS

Quail Map Company

Each volume of the Quail reference work contains detailed track diagrams for the former British Rail Regions, plus private railways and industrial layouts. Includes extensive notation, list of abbreviations, engineers line references and an index to stations, lines and selected other places. Used extensively throughout the railway industry, the following volumes are currently available:

Railway Track Diagrams 1: Scotland and the Isle of Man £7.50
Railway Track Diagrams 3: Great Western £7.50
Railway Track Diagrams 5: England South & London Underground £8.95
Railway Track Diagrams 7: New South Wales Metropolitan Areas £6.50

Please Note: Numbers 2, 4 & 6 in the series are completely out of print. If you would like to be notified when new titles in this series become available, please contact our Mail Order Department. Alternatively, please see our advertisements in Today's Railways or Entrain magazines for up to date publication information.

Please add postage: 10% UK, 20% Europe, 30% Rest of World.

3. SERVICE DMUS

This section lists vehicles not used for passenger-carrying purposes. Some vehicles are numbered in the special service stock number series.

CLASS 114/1   ROUTE LEARNING UNIT

DMB–DT. Converted 1992 from Class 114/1. Gangwayed within unit.

Construction: Steel.
Engines: Two Leyland TL11/40 of 153 kW (205 h.p.) at 1950 r.p.m.
Transmission: Mechanical. Cardan shaft and freewheel to a four-speed epicyclic gearbox with a further cardan shaft to the final drive, each engine driving the inner axle of one bogie.
Maximum Speed: 70 m.p.h.   **Couplings:** Screw.
Bogies: DD9 + DT9.   **Multiple Working:** Blue Square.
Brakes: Twin pipe vacuum.   **Dimensions:** 20.45 x 2.82 m.
Doors: Manually operated slam/roller shutter.
Non-Standard Livery: Grey, red and yellow.

977775. DMB. Dia. DZ518. Lot No. 30209 Derby 1957. 39.0 t.
977776. DT. Dia. DZ516. Lot No. 30210 Derby 1957. 29.2 t.

| - | | **0** | E | | TE | 977775 | (55928) | 977776 | (54904) |

CLASS 122   ROUTE LEARNING UNIT

DM. Converted 1995 from DMBS. Non gangwayed single car with cab at each end.

Construction: Steel.
Engines: Two Leyland 1595 of 112 kW (150 h.p.) at 1800 r.p.m.
Transmission: Mechanical. Cardan shaft and freewheel to a four-speed epicyclic gearbox with a further cardan shaft to the final drive, each engine driving the inner axle of one bogie.
Maximum Speed: 75 m.p.h.   **Couplings:** Screw.
Bogies: DD10.   **Multiple Working:** Blue Square.
Brakes: Twin pipe vacuum.   **Dimensions:** 20.45 x 2.82 m.
Doors: Manually operated slam.

55012. DM. Dia. DX202. Lot No. 30419 Gloucester 1958. Converted by ABB Doncaster 1995. 36.5 t.

| - | **LH** | E | E | TE | 55012 |

CLASS 930   SANDITE/DE-ICING UNIT

DMB–T–DMB. Converted 1993 from Class 205. Gangwayed within unit. Sandite trailer 977870 is replaced by de-icing trailer 977364 as required.

Construction: Steel.
Engine: One English Electric 4SRKT Mk. 2 of 450 kW (600 h.p.) at 850 r.p.m.
Transmission: Electric. Two English Electric EE507 traction motors mounted on the bogie at the non-driving end of each power car.
Maximum Speed: 75 m.p.h.   **Bogies:** SR Mk. 4.
Brakes: Electro-pneumatic and automatic air.
Doors: Manually operated slam.   **Couplings:** Drophead buckeye.
Multiple Working: Classes 201–207.
Dimensions: 20.33 x 2.82 x 3.87 m. (DMB); 20.28 x 2.82 m.

977939–977940. DMB. Dia. DZ537. Lot No. 30671 Eastleigh 1962. 56.0 t.
977870. T. Dia. DZ533. Lot No. 30542 Eastleigh 1960. 30.5 t.

930 301	**RO**	NR	*NR*	SU	977939	(60145)	977870	(60660)
					977940	(60149)		

CLASS 960 ULTRASONIC TESTING/TRACTOR UNIT

DM–DM. Converted 1986 from Class 101. Gangwayed within unit. Often operates with either 975091 or 999602 as a centre car.

For detail see page 10 except:

Brakes: Air.

977391. DM. Dia. DZ503. Lot No. 30500 Metro-Cammell. 1959. 32.5 t.
977392. DM. Dia. DZ503. Lot No. 30254 Metro-Cammell. 1956. 32.5 t.

	RK	NR	*SO*	RG	977391	(51433)	977392	(53167)

CLASS 960   TEST UNIT (Iris 2)

DM–DM. Converted 1991 from Class 101. Gangwayed within unit.

For details see page 10.

977693. DM. Dia. DZ503. Lot No. 30261 Metro-Cammell. 1957. 32.5 t.
977694. DM. Dia. DZ503. Lot No. 30276 Metro-Cammell. 1958. 32.5 t.

-	**SO**	NR	*SO*	BY	977693	(53222)	977694	(53338)

CLASS 960   SANDITE UNIT

DMB. Converted 1991/93 from Class 121. Non gangwayed.

For details see page 11.

977722-977723. DMB. Dia. DX515. Lot No. 30518 Pressed Steel 1960. 38.0 t.
977858–60/66/73. DMB. Dia. DZ526. Lot No. 30518 Pressed Steel 1960. 38.0 t.

960 010	M	NR	NR	AL	977858	(55024)
960 011	RK	NR	NR	AF	977859	(55025)
960 012	N	NR	NR	AF	977860	(55022)
960 013	RO	NR	NR	AL	977866	(55030)
960 014	RK	NR	CR	AL	977873	(55028)
960 021	RO	NR	NR	AL	977723	(55021)

CLASS 960 — EMERGENCY TRAIN UNITS

Under conversion from Class 121 for Severn Tunnel emergency train. For details see page 11.

| - | | NR | NR | CP | 55027 |
| - | | NR | NR | CP | 55031 |

CLASS 960 — TRACK ASSESSMENT UNIT

Converted from Class 121. For details see above.

Non-standard Livery: Yellow with blue & purple logo.

| 977968 | 0 | GT | GT | Rugby Rail Plant Depot | 55029 |

CLASS 960 — SANDITE UNIT

DMB. Converted 1991 from Class 122. Non gangwayed.

Construction: Steel.
Engines: Two Leyland 1595 of 112 kW (150 h.p.) at 1800 r.p.m.
Transmission: Mechanical. Cardan shaft and freewheel to a four-speed epicyclic gearbox with a further cardan shaft to the final drive, each engine driving the inner axle of one bogie.
Maximum Speed: 70 m.p.h.
Bogies: DD10.
Brakes: Twin pipe vacuum.
Doors: Manually operated slam.
Couplings: Screw.
Multiple Working: Blue Square.
Dimensions: 20.45 x 2.82 m.

975042. DMB. Dia. DX516. Lot No. 30419 Gloucester 1958. 36.5 t.

| 960 015 | RO | NR | NR | AL | 975042 | (55019) |

UNCLASSIFIED — DE-ICING UNIT

T. Converted 1960 from 4-Sub EMU vehicle. Non gangwayed. Operates with 977939/40.

Construction: Steel.
Maximum Speed: 70 m.p.h.
Bogies: Central 43 inch.
Brakes: Electro-pneumatic and automatic air.
Doors: Manually operated slam.
Couplings: Drophead buckeye.
Multiple Working: SR system.
Dimensions:

977364. T. Dia. EZ520. Southern Railway Eastleigh 1946. 29.0 t.

| - | | RO | NR | NR | SU | 977364 | (10400) |

UNCLASSIFIED   TRACK ASSESSMENT UNIT

DM–DM. Purpose built service unit. Gangwayed within unit.

Construction: Steel.
Engine: One Cummins NT-855-RT5 of 213 kW (285 h.p.) at 2100 r.p.m. per power car.
Transmission: Hydraulic. Voith T211r with cardan shafts to Gmeinder GM190 final drive.
Maximum Speed: 75 m.p.h.   **Couplers:** BSI automatic.
Bogies: BP38 (powered), BT38 (non-powered).
Brakes: Electro-pneumatic.   **Dimensions:** 20.06 x 2.82 m.
Doors: Manually operated slam & power operated sliding.
Multiple Working: Classes 142, 143, 144, 150, 153, 155, 156, 158, 159, 170.

999600. DM. Dia. DZ536. Lot No. 4060 BREL York 1987. 36.5 t.
999601. DM. Dia. DZ536. Lot No. 4061 BREL York 1987. 36.5 t.

-	RK	NR	SO	ZA	999600 999601

UNCLASSIFIED   ULTRASONIC TEST UNIT

T. Converted 1986 from Class 432 EMU. Gangwayed. Operates with 977391/2

Construction: Steel.   **Maximum Speed:** 70 m.p.h.
Bogies: SR Mk. 6.   **Couplings:** Screw.
Brakes: Twin pipe vacuum.   **Multiple Working:** Blue Square.
Doors: Manually operated slam.   **Dimensions:** 19.66 x 2.82 m.

999602. T. Dia. DZ531. Lot No. 30862 York 1974. 55.5 t.

-	SO	SO	SO	ZA	999602 (62483)

UNCLASSIFIED TRACK ASSESSMENT/RECORDING UNIT

DM–DM. Universal track recording unit for video inspections, for measuring rail profiles etc.. Full details awaited.

Construction: **Engine:**
Transmission:
Maximum Speed: 100 m.p.h.   **Weight:** 70 t.
Brakes: **Dimensions:** 20.06 x 2.82 m.

999700. DM.
999701. DM.

Non-standard Livery: Yellow with blue & purple logo.

-	0	GT	GT	999700 999701

4. DMUS AWAITING DISPOSAL

The list below comprises vehicles awaiting disposal which are stored on the Railtrack network, together with vehicles stored at other locations (e.g. repair facilites) which, although awaiting disposal, remain Railtrack registered. This includes vehicles for which sales have been agreed, but collection by the new owner had not been made at the time of going to press.

Class 100

977191	B	?	ZC			

Class 101

101 654	RR	A	PY	51800	54408	
101 656	RR	A	LO	51230	54056	
101 657	RR	A	LO	51175	53211	54085
101 660	RR	A	PY	51213	54343	
101 665	RR	A	PY	51429	54393	
101 681	RR	A	PY	51228	51506	
101 682	RR	A	PY	53256	51505	
101 683	RR	A	LO	51177	53269	
101 684	S	A	PY	51187	51509	
101 686	S	A	PY	51231	51500	
101 687	S	A	PY	51247	51512	
101 690	S	A	PY	51435	53177	
101 691	S	A	PY	51253	53171	
101 694	S	A	PY	51188	53268	
101 695	S	A	PY	51226	51499	
101 835	RR	A	PY	51432	51498	
101 840	N	A	LO	53311	53322	
960 991	N	NR	DY	977895	(53308)	977896	(53331)
960 992	BG	NR	DY	977897	(53203)	977898	(53193)
960 993	BG	NR	DY	977899	(51427)	977900	(53321)
960 994	BG	NR	DY	977901	(53200)	977902	(53231)
960 995	BG	NR	DY	977903	(53208)	977904	(53291)

Spare:	RR	A	LO	51442	51533	54091		
Spare:	RR	A	PY	51189	51213	51442	51463	51496
Spare:	RR	A	PY	54055	54061	54062	54347	54352
Spare:	RR	A	PY	54343	54358	54365		
Spare:	RR	A	ZH	51185	51224	51428	53163	
Spare:	BG	A	NL	54342				
Spare:	BG	A	PY	54350				
Spare:	RR	A	ZH	51213				
Spare	RR	A	BP	59303				
Spare	G	A	BP	59539				

Class 117

117 301	RR	A	PY	51353	51395
117 306	RR	A	AL	51369	51411
117 308	RR	A	PY	51371	51413

117 310	**RR**	A	PY	51373	59486	51381
117 313	**RR**	A	PY	51339		51382
117 701	**N**	A	PY	51350		51392
117 702	**N**	A	PY	51356		51398
117 704	**N**	A	PY	51341		51383
117 706	**N**	A	PY	51366		51408
117 707	**N**	A	PY	51335		51377
Spare	**N**	A	PY	51358	51411	
Spare	**RR**	A	PY	59492	59500	59509	59521

Class 141

141 101	**WY**	CD	ZF	55501	55521
141 105	**WY**	CD	ZF	55505	55525
141 106	**WY**	CD	ZF	55506	55526
141 112	**WY**	CD	ZF	55512	55532
141 116	**WY**	CD	MM	55516	55536
141 118	**SO**	CD	ZF	55518	55538

Note: 141 101, 141 105 and 141 116 are pending sale to Iran.

Class 951

977696	**N**	NR	ZG

5. CODES

5.1. LIVERY CODES

Code   *Description*

AL   Advertising livery (see class heading for details).
AN   Anglia Railways Class 170s (white & turquoise with blue vignette).
AR   Anglia Railways (turquoise blue with a white stripe).
AV   Arriva Trains Northern (turquoise blue with white doors).
B   BR blue.
BG   BR blue & grey lined out in white.
BI   Visit Bristol promotional livery (deep blue with various images).
BL   BR blue with yellow cabs, grey roof, large numbers & BR logo..
CO   Centro (grey/green with light blue, white & yellow stripes).
CR   Chiltern Railways (blue & white with a thin red stripe).
CT   Central Trains (two-tone green with yellow doors. Blue flash and red strpe at vehicle ends.).
CX   Connex (white with yellow lower body & blue solebar).
DC   Scenic lines of Devon & Cornwall promotional livery (black with gold cantrail stripe).
FG   First Group corporate Inter-City livery (indigo blue with a white roof & gold, pink & white stripes).
FS   First Group corporate regional/suburban livery (indigo blue with pink & white stripes).
G   BR Southern Region/SR or BR DMU (green with straw stripe on coaching stock).
GM   Greater Manchester PTE (light grey/dark grey with red & white stripes).
HW   Heart of Wales Line promotional livery (orange with yellow doors).
LH   BR Loadhaul (black with orange cabsides).
M   BR maroon (Maroon lined out in straw & black).
MM   Midland Mainline (Teal green with cream lower body sides & three orange stripes).
MN   New Midland Mainline Meridian (details to follow).
MT   Old Merseytravel (yellow/white with grey & black stripes).
MY   New Merseytravel (yellow/white with grey stripe).
N   BR Network South East (white & blue with red lower bodyside stripe, grey solebar).
NS   Northern Spirit/Arriva Trains Northern (turquoise blue with lime green 'N').
NT   BR Network South East (white & blue with red lower bodyside & cantrail stripes).
NW   North West Trains/First North Western (blue with gold cantrail stripe & star).
O   Non standard livery (see class heading for details).
P   Porterbrook Leasing Company (purple & grey or white).
PS   Provincial (dark blue/grey with light blue & white stripes).
RE   Provincial/Regional Railways Express (light grey/buff/dark grey with white, dark blue & light blue stripes).
RK   Railtrack (green & blue).

RN	North West Regional Railways (dark blue/grey with green & white stripes).
RO	Old Railtrack (orange with white & grey stripes).
RR	Regional Railways (dark blue/grey with light blue & white stripes, three narrow dark blue stripes at vehicle ends).
S	Old Strathclyde PTE (orange & black lined out in white).
SC	New Strathclyde PTE (carmine & cream lined out in black & gold).
SL	Silverlink (indigo blue with white stripe, green lower body & yellow doors).
SN	Govia South Central (white & dark green with light green patch near cab ends).
SO	Serco Railtest (red/grey).
SP	New Strathclyde PTE Class 170/334 livery (carmine & cream, with a turquoise stripe).
SR	ScotRail (white, terracotta, purple & aquamarine).
SW	South West Trains (white & dark blue with black window surounds, red doors & red panel with orange stripe at unit ends).
TT	Thames Trains (blue with lime green doors).
TW	Tyne & Wear PTE (white & yellow with blue stripe).
TX	Arriva Trans-Pennine Express (plum with yellow 'N').
VL	Valley Lines (dark green & red with white & light green stripes. Light green doors).
VT	New Virgin Trains (silver, with black window surrounds, white cantrail stripe & red roof. Red swept down at unit ends. Black and white striped doors).
VW	Visit Wales promotional livery (green & red with various images).
WB	Wales & Borders Alphaline (metallic silver with blue doors).
WX	Heart of Wessex Line promotional livery (red with yellow doors).
WT	Wessex Trains Alphaline (metallic silver with maroon doors).
WY	Old West Yorkshire PTE (red/cream with thin yellow stripe).
WZ	Wessex Trains claret promotional livery with various images.
YN	West Yorkshire PTE (red with light grey 'N').
YP	New West Yorkshire PTE (red with grey semi-circles).

5.2. OWNER CODES

A	Angel Train Contracts.
BC	Bridgend County Borough Council.
CD	Cotswold Rail Ltd.
E	English Welsh & Scottish Railway.
H	HSBC Rail (UK) Ltd.
GT	GTRM.
HD	Hastings Diesels Ltd.
HX	Halifax Asset Finance Ltd.
NR	Network Rail.
P	Porterbrook Leasing Company.
RD	Rhondda Cynon Taff District Council.
SO	Serco Railtest.
W	Wiltshire Leasing Co. (First Group).

5.3. OPERATION CODES

AR	Anglia Railways.
AV	Arriva Trains Northern.
CR	Chiltern Railways.
CT	Central Trains.
E	English Welsh & Scottish Railway.
GT	GTRM.
GW	First Great Western.
MM	Midland Mainline.
NR	Network Rail.
NW	First North Western.
SC	South Central.
SL	Silverlink.
SO	Serco Railtest.
SR	ScotRail.
SS	Normally used only on special or charter services.
SW	South West Trains.
TT	Thames Trains.
VX	Virgin Cross Country
VL	Wales & Borders Trains (Valley lines business unit).
WB	Wales & Borders Trains
WX	Wessex Trains.

5.4. ALLOCATION & LOCATION CODES

Code	Location	Operator
AF	Ashford Chart Leacon T&RSMD (Kent)	Bombardier Transportation
AL	Aylesbury TMD	Chiltern Railways
BP	Blackpool North CS	*Storage location only*
BY	Bletchley T&RSMD	Silverlink
CF	Cardiff Canton TMD	EWS
CH	Chester TMD	First North Western
CK	Corkerhill TMD (Glasgow)	ScotRail
CZ	Central Rivers T&RSMD (Barton-under-Needwood, near Burton-on-Trent)	Bombardier Transportation
DY	Derby Etches Park T&RSMD	Maintrain
HA	Haymarket TMD (Edinburgh)	ScotRail
HT	Heaton T&RSMD (Newcastle)	Arriva Trains Northern
LO	Longsight Diesel (Manchester) TMD	First North Western
MM	Fire Service College Moreton-in-Marsh	Cotswold Rail
NC	Norwich Crown Point T&RSMD	Anglia Railways
NH	Newton Heath TMD (Manchester)	First North Western
NL	Neville Hill T&RSMD (Leeds)	Arriva Trains Northern
OM	Old Oak Common CARMD (London)	First Great Western
PY	MoD DERA Pigs Bay (Shoeburyness)	MoD Defence Rail Group
RG	Reading TMD	Thames Trains
SA	Salisbury TMD	South West Trains
SE	St. Leonards TMD (Hastings)	St. Leonards Railway Engineering
SU	Selhurst T&RSMD (Croydon)	South Central
TS	Tyseley T&RSMD (Birmingham)	Maintrain
ZA	Railway Technical Centre Derby	Serco Railtest/AEA Technology
ZC	Crewe Works	Bombardier Transportation
ZG	Eastleigh Works	Alstom
ZH	Springburn Works Glasgow	Railcare

ABBREVIATIONS

CARMD	Carriage Maintenance Depot
DERA	Defence Evaluation & Research Agency
T&RSMD	Traction and Rolling Stock Maintenance Depot.
TMD	Traction Maintenance Depot.
TMD (D)	Traction Maintenance Depot (Diesel).
SD	Servicing Depot.